DATSUN Z SERIES
The Complete Story

Other Titles in the Crowood AutoClassics Series

AC Cobra	Brian Laban
Aston Martin: DB4, DB5 and DB6	Jonathan Wood
Aston Martin and Lagonda V-Engined Cars	David G Styles
Austin-Healey 100 and 3000 Series	Graham Robson
BMW M-Series	Alan Henry
The Big Jaguars	Graham Robson
Ford Capri	Mike Taylor
The Big Jaguars: 3½-Litre to 420G	Graham Robson
Jaguar E-Type	Jonathan Wood
Jaguar XJ Series	Graham Robson
Jaguar XK Series	Jeremy Boyce
Jaguar Mk1 and 2	James Taylor
Jaguar S-Type and 420	James Taylor
Lamborghini Countach	Peter Dron
Land Rover	John Tipler
Lotus and Caterham Seven	John Tipler
Lotus Elan	Mike Taylor
Lotus Esprit	Jeremy Walton
Mercedes SL Series	Brian Laban
MGA	David G Styles
MGB	Brian Laban
Morgan: The Cars and The Factory	John Tipler
Porsche 911	David Vivian
Porsche 924/928/944/968	David Vivian
Range Rover	James Taylor and Nick Dimbleby
Sprites and Midgets	Anders Ditlev Clausager
Sunbeam Alpine and Tiger	Graham Robson
Triumph TRs	Graham Robson
Triumph 2000 and 2.5PI	Graham Robson
TVR	John Tipler
VW Beetle	Robert Davies
VW Golf	James Ruppert

Datsun Z
SERIES

THE COMPLETE STORY

David G. Styles

First published in 1996 by
The Crowood Press Ltd
Ramsbury, Marlborough
Wiltshire SN8 2HR

© David G Styles 1996

The right of Dr David G Styles to be identified as the author of this work is asserted in accordance with the Copyright, Design and Patents Act 1988

All rights reserved. No part of this publication may be reproduced or transmitted in any form or by any means, electronic or mechanical, including photocopy, recording, or any information storage and retrieval system, without permission in writing from the publishers.

British Library Cataloguing-in-Publication Data
A catalogue record for this book is available from the British Library.

ISBN 1 86126 001 6

Picture Credits
The photographs in this book are from the author's collection or have been obtained from the Nissan Motor Company Limited in Tokyo, Nissan Europe NV, Nissan UK Limited, the National Motor Museum at Beaulieu, Bill Piggot, Barney Sharratt, Len Welch, *Road and Track* and Toyota Motor GB Limited. Road test data is reproduced by kind permission of the editors of *Road and Track* and *Autocar and Motor* and drawings are reproduced from the parts and works manuals published by the Nissan Motor Company Limited, with the help of Matthew Humphris, Director of Humphris of Oxford Limited.

Dedication
To my family – my wife Ann, my daughter Emma and my son Philip – all of whom have quietly tolerated my writings and the long hours of sitting at a word processor or in a library somewhere. Without their indulgence, I could not write.

Typeface used: New Century Schoolbook.

Typeset and designed by
D&N Publishing
Ramsbury, Wiltshire

Printed and bound by The Bath Press.

Contents

	Acknowledgements	6
	Introduction	7
	Evolution	8
1	From Little Acorns	10
2	Creating a Sports Car, Creating an Industry	18
3	The Path to Success and to the Z	28
4	The Z Age Begins: the 240Z and the Fairlady Z	42
5	The 240Z in the Market	55
6	Development of the 240Z – and the 260Z Arrives	68
7	Extending the Line – the 280Z and the 280ZX	83
8	New Age, New Heights – the 300ZX	101
9	Facing the Competition	123
10	Road Tests and Press Reviews	146
11	Proving the Z in Racing and Rallying	165
12	Running and Restoring the Z Series	177
	Further Reading	189
	Index	190

Acknowledgements

Many people made a contribution to this book. The first person, without doubt, to whom I must express most sincere thanks for allowing me to make a thorough pest of myself in my quest for fill-in facts and early history is Julia Smith, Nissan Europe's Assistant Manager of Public and Corporate Affairs in Amsterdam and her assistant, Megumi Ishida. Without the interest shown by Julia and her willingness to keep asking for data, the book would not have been finished and certainly would not have the detail that it does.

Toyokazu Ishida, Manager, Planning and Administration Section at the Nissan Motor Company Limited in Tokyo, fed quite a lot of material to me through Julia, including some fascinating early photographs and much technical data. Still with Nissan, Leslie Alder at Nissan UK's factory in Washington put me in contact with Linda Robinson and Mike Parkinson at their offices in Rickmansworth. The result was the provision of quite a bit of useful material on the later 300ZX models.

Much of the material provided by Toyokazu Ishida was printed in Japanese, so I had to find someone to translate it. I was privileged to come into contact with two postgraduate students at Loughborough University: Mamoru Hirata and Kazufumi Nagashima kindly translated the text to English, which was of immense help in plotting the early history of Nissan and its antecedents. I am truly grateful to them and hope that the finished work comes up to their hopes and expectations.

In my search for pictures, Len Welch comes high on the list of the knowledgeable and helpful. Many of the pictures here were obtained from his collection. The material on the Toyota 2000GT came from Ginny McDonald of Toyota Motor Manufacturing (UK) Limited. Road test data is reproduced with kind permission of *Autocar and Motor* in Britain, as well as *Road and Track* in the United States. To the editors of both those journals I express my thanks. Finally, my thanks go to all the other people who have given help in various ways, and to Z enthusiasts the world over for giving me the reason to write this book.

DAVID G STYLES

Introduction

This book is a significant departure for me and for my publisher, for it is the first book about Japanese sports cars for both of us. I have long thought that the Datsun Z models were worth a second look. When I first rode in a 240Z in the early 1970s I was impressed with its performance, likening it to the then recently demised Austin Healey 3000. I could see that the 240Z might become a worthy successor to the Austin Healey, but like many people I was not sure the Datsun would last the course; Japanese cars had a very small share of the British market.

As the 240Z became the 260Z in Britain, and the 280Z then arrived in the United States, followed by the 2+2s, it began to dawn on a lot of people that the Z models were here to stay; production volumes by the end of 1975 made it quite plain that Nissan had a winner. As the 280ZX superseded the first generation Z, the upward move in the market had begun and Nissan made a transition from Sports/Grand Tourers to Grand Tourers/Sports cars, with a strong emphasis on the 'Grand'.

At times it seemed that Nissan had lost its way in its quest for a super affordable sports car, but then came the 300ZX. Unashamedly luxurious, the 300ZX developed into the Z-32 version, the new 300ZX, and in the process recovered much of the charisma of the original 240Z, albeit on a higher price plane. But the Z has become a classic and there is a growing band of sporting classic enthusiasts who are making the acquaintance of the Datsun and Nissan Z models because they were well designed and well made and are fun to drive. With this series the Nissan Motor Company brought sports car motoring to many who would otherwise never have been able to afford the experience.

Evolution

1912	Kwaishinsha Motor Company first formed, to make DAT cars
1926	Kwaishinsha and Jitsuyo Jidosha merge to form DAT Jidosha Seizo
1931	DAT Jidosha Seizo Company affiliates with Tobata Foundry Company
1933	Jidosha Seizo Kabushiki Kaisha Company formed from DAT and Tobata
1934	Corporate name changed to Nissan Motor Company Limited
1936	Nissan purchases plant and designs from Graham Paige
1939	Manufacture of military vehicles begins
1945	Post-war reconstruction of Japanese industry begins
1952	First ever sports car in Japan announced – the Datsun CD-3
1953	Nissan enters licence agreement with Austin Motor Company
1958	Nissan 210 wins class in Mobilgas Australia Rally
1959	Datsun Model S211 introduced
1960	Agreement with Austin Motor Company expires
1960	SP212 model goes on sale
1961	Fairlady 1500 Sports (SP310) launched at Tokyo Motor Show
1962	New Nissan Oppama factory opens
1964	Fairlady 1600 and 2000 (SP311) succeed SP310 at Tokyo Show
1965	Nissan Zama plant completed and opened
1966	Nissan and Prince Motor Company merge
1967	Model SR311 Fairlady takes place of SP311
1968	Hardtop version of SR311 launched in Japan
1969	First Z announced – known as PS30/Z432 or HS30/240Z
1970	240Z makes international rally debut in RAC Rally – finished 7th
1971	Monte Carlo Rally – 240Zs in 5th and 10th places
1971	East Africa Safari Rally – 240Zs 1st, 2nd, 7th and team prize
1972	Monte Carlo Rally – 240Z 3rd place
1972	Montana Rally (Portugal), Kenya 2000 Rally – 240Z 1st
1973	East Africa Safari Rally, Tanzania 1000 Rally – 240Z 1st
1974	260Z with Model L26 engine of 2,565cc replaces 240Z
1974	2+2 version of 260Z launched to join two-seater version
1975	280Z and 280Z 2+2 (2,753cc) introduced to meet United States emission standards
1978	280ZX and 280ZX 2+2 announced
1981	Agreement entered into with Alfa Romeo to produce ARNA
1982	New 280Z Turbo 2+2 released
1983	ARNA introduced in Europe, forms basis of Cherry Europe model

1983	Model 300ZX launched with VG30 V-6 engine
1987	Concept cars at Tokyo Motor Show, modelling new 300ZX line
1989	New generation 300ZX with VG30DE/VG30DETT V-6 engines announced
1990	Nissan 300ZX Turbo wins IMSA GTO Race at Miami
1992	300ZX convertible announced with non-turbocharged engine only
1992	Nissan 300ZX Turbo wins IMSA GTU Championship
1993	300ZX Turbo wins Miami Grand Prix and Sebring 12 Hours Race
1994	Nissan 300ZX Turbo wins Daytona 24 Hours Race
1994	300ZX withdrawn from British market

1 From Little Acorns

Japan is today a constitutional monarchy, consisting of four main islands: Kyushu and Shikoku to the south, Honshu in the middle and Hokkaido to the north. The total archipelago consists of over a thousand islands, bordered by the Sea of Japan, the Sea of Okhotsk and the Pacific Ocean. The land mass of Japan is roughly half as big again as that of the British Isles, but it supports over twice as many people. The central island of Honshu is the most industrialized and it is on this island that the capital, Tokyo, is located.

As an absolute monarchy, with its law and order dictated by the Shogun, Imperial Japan closed its doors to the outside world in 1637. As the pressures of world trade brought the United States to force Japan to open its doors again in 1853, so the move to constitutionalism began and the last Shogun abdicated in 1867, restoring executive power to the Emperor. Over thirty more years, a constitutional code of law was established with a formal constitution being established in 1889.

So when Carl Benz patented his Motor Wagen in 1886, Japan was still a relatively closed country, with very little outside intrusion. Just before the turn of the 20th Century, the nobility of Japan had been made aware of the existence of the automobile – but as in so many western nations, those in power were a long way from seeing the true potential of motor vehicles in their society.

With the Imperial shutters kept down for so long, it is hardly surprising that technology took a long time to penetrate Japanese shores. It may be said that at the time they did not need it, but once they were aware of the products of modern science, the Japanese became avid students of things mechanical and electrical. They had their own science, of course, and would certainly have reached equality with the West eventually,

This outline map of Japan shows the four main islands and the myriad of little islands that make up the archipelago.

From Little Acorns

but on the principle of 'why re-invent the wheel?' the Japanese acquired a reputation for being experts at imitation. The march of time continued, the automobile became accepted as a fact of life and it found its way into Japan. It could not be long before the natural sense of Japanese pride would drive someone to initiate the manufacture of motor cars in their own land.

THE PIONEERING EFFORTS OF MASUJIRO HASHIMOTO

Masujiro Hashimoto was a young Japanese mechanical engineer who had trained in his homeland and then the United States. Having seen the benefits that the automobile had brought to that country, he was eager to see that his own should gain the advantage of motorized transport too. Hashimoto was not an especially wealthy man, however, though he must have been well connected, for such was the system in the Japan of those days that the situation that brought him into the motor manufacturing business could not have happened without his having influential connections.

Three men were known to Hashimoto as possible backers and so he approached them all. When all three had agreed to give him support, he established his new business, in 1911, with the three as shareholders. Those three men were Kenjiro Den, Rokuro Aoyama and Metaro Takeuchi. In typically honourable fashion, as an expression of his gratitude to his backers for their support, Hashimoto used the initial letter of each of his backers' family names as the name for his new product. So was born the Japanese motor manufacturer Kwaishinsha and its product was the DAT car. The word formed from the letters DAT meant 'hare', which was fortuitous as it gave the cars the name of an animal of speed.

The Kwaishinsha factory was established in the Azabu-Hiroo district of Tokyo. Hashimoto was a pioneer of the automobile in Japan and was determined to see an automobile industry blossom there. He already believed that industry was the way to feed Japan's growing population, and that the

The first Kwaishinsha-built car, an open four-seater built in 1914 and known as the DAT, so named in honour of the financial backers of the original company.

From Little Acorns

establishment of an international economy would create growth on a scale that would be impossible to achieve on the agricultural base that existed at that time. For twenty years Hashimoto was the only person in Japan to operate a motor business. His first car rolled out of the Azabu-Hiroo works during 1914 and it established the line that ultimately became the Nissan Motor Company Limited.

Japan had formed an alliance with Britain in 1902, as part of the strategy to protect its territorial acquisitions of the southern half of Sakhalin island from the Russians and Korea from the Chinese. World War One put the whole world at risk and drew Japan into the conflict, though with minimal involvement. The major benefit to Japan of its alliance at that time with Britain was the assurance of protection for its territories and, in direct benefit of being 'on the right side' after the war was over, Japan acquired sovereignty over the German-held islands in the North Pacific, including Tinian. During the war itself the manufacture of motor cars was limited.

THE JITSUYO JIDOSHA COMPANY AND THE MERGER WITH KWAISHINSHA

Jitsuyo Jidosha was formed in 1919, after the end of World War One and during a period of industrial growth for Japan. Its initial purpose was to manufacture three wheeled vehicles designed by an American engineer named William R. Gorham. Gorham had visited the Far East and had seen the need for a small, economical vehicle that could ply the narrow streets of Japanese and Chinese cities. So he developed his tricycle, which was taken up under licence by Jitsuyo Jidosha and put into production in their Osaka factory, the vehicles being sold as taxis.

The manufacture of Gorham tricycles was remarkably efficient, with Jitsuyo Jidosha being almost certainly the first motor manufacturer in Japan to establish any kind of production line. Also, they set up a machine shop to make components and assembly jigs, so that line manufacture was soon a practice in the infant Japanese motor manufacturing industry. Within a year, Gorham came up with a new design of his vehicle and added a wheel to it to make a light car, the Lila, which went into production in 1921. A key to the success of both the tricycle and the four-wheeled car was that Jitsuyo Jidosha established their production method on the basis of ensuring that components were interchangeable as far as possible. Many manufacturers in the western hemisphere had still not developed that technique.

By the mid 1920s the paths of the Kwaishinsha Motor Company and Jitsuyo-Jidosha had crossed and Hashimoto was proposing that the two companies should join forces, as the interest in motor cars was growing among Japan's well-to-do families and there was business to be done. Hashimoto had seen the Jitsuyo factory and admired the production line techniques, as well as the principle of using interchangeable parts to reduce overall manufacturing costs. He had designed a military armoured car (the Type 41) and gained official certification for the design, but the finances of the Kwaishinsha company were now becoming hard pressed. So these two companies merged in 1926 to create the DAT Jidosha Seizo Company Limited, with a new factory at Osaka. Over the next few years, a series of military vehicles – armoured cars, personnel carriers and trucks – was produced. By 1930, a reorganization took place and the directors decided a new name was needed for their cars; the name DATSON was chosen, being the 'son of DAT'. The spelling of the name was later changed to 'sun' to make the name look better for the English-speaking export market.

THE TOBATA IMONO COMPANY AND THE BIRTH OF NISSAN

Yoshisuke Ayukawa was an entrepreneur and president of his own foundry company called Tobata Imono. As his company and his capital grew, he developed an ambition to mass produce cars in the fashion of the growing American motor industry. In June 1931, he realized the first step of that ambition by acquiring the assets and shares of DAT Jidosha Seizo, making the company the automobile division of Tobata Imono.

Two years later, on 26 December 1933, three days after the birth of Crown Prince Akihito, the Tobata Imono Automobile Division was separated from the parent company, established as an independent company under the name of Jidosha Seizo Kabushiki Kaisha and relocated to a new 32 acre (132,000 square metre) site at Yokohama. The business world of Tokyo forecast that the new Yokohama plant would be overgrown with weeds inside two years, as no-one believed that mass production of cars would ever come to Japan.

Funding for the new firm came from a holding company managed by Yoshisuke

---

Yoshisuke Ayukawa – Founder of the Nissan Motor Company

Yoshisuke Ayukawa visited the United States in 1906, staying with an American family while he studied. Exactly where he stayed, or with whom, and where he studied is not available to us now, though a photograph survives that shows him sitting on the steps of his host family's home. The group consists of two men, presumably father and son, standing alongside three seated women and a young girl, with young Ayukawa seated in the middle of the group.

During his stay in the United States, young Ayukawa was very much impressed with several aspects of American life, not least the presence of the automobile. His host owned a car and clearly the young Japanese guest was frequently taken out in it. The standard of living witnessed by Ayukawa persuaded him that industry in Japan was the way ahead, for by exporting manufactured goods, the nation's prosperity and the standard of living of its people could improve considerably.

It is apparent that Ayukawa came from a quite well-to-do family in Japan, for unless he did it is highly unlikely that he would have been able to study in the United States, or even visit that country, and he certainly would not have been able to undertake the business venture he embarked upon in 1911 after his return. The Tobata Imono Company was a foundry established to produce high quality castings, and he incorporated into that business all the lessons and techniques he had acquired from his American education. His company was soon turning out metal products equal to those of Europe and the United States and Tobata castings were among the first exports from Japan.

By 1930 Ayukawa was looking for ways of expanding his business into other related fields. Fortunately for him, the Jidosha Seizo DAT Company were looking for investment capital to progress the next Datson car into production. Tobata Imono came to the rescue with the aid of the Nihon Sangyo Bank and the Jidosha Seizo Kabushiki Kaisha Company was born. May 1934 saw the change of name to the Nissan Motor Company Limited. The first stage in Yoshisuke Ayukawa's dream had come true and in that first year the new company built ten cars. By 1935, the number had risen to 3,800 and Nissan cars, built in Yokohama, were being exported to Australia.

From Little Acorns

Ayukawa, called Nihon Sangyo. That holding company was already known in Tokyo's financial circles as 'Ni-San'. By May 1934, Nihon Sangyo acquired the remaining share stock of Jidosha Seizo Kabushiki Kaisha to become the sole shareholder. The name of the motor company was now changed to Nissan Motor Company Limited. Through all these corporate changes, the product name of the motor company remained 'Datsun' and it was to continue to be made under that name for many more years. One of the major benefits of the merger was that the Tobata Imono company brought with it a contract for the manufacture of spare parts for Ford and Chevrolet cars, which was now transferred to the Osaka plant, while the Yokohama factory was extended and developed for increased production of Datsun cars.

By 1935, the Datsun product was firmly established and Types 10 to 14 had already been marketed. But now, the first small passenger car rolled off the line at Yokohama and an export link with Australia was established. Datsun cars were the very symbol of Japanese industrial advances of the time and the slogan 'The Rising Sun as the flag and Datsun as the car of choice' was adopted by Nissan. With such a late start in the motor industry, the advancement of the Nissan Motor Company Limited was little short of remarkable. To be able to export to Australia and compete in that market with British cars was no mean achievement either.

One of Nissan's domestic successes was the Roadster two-seater, which was essentially a crib of Britain's renowned Austin Seven, but in 1936 a link was forged with one of America's fading greats, Graham-Paige. One of Graham-Paige's claims to greatness was the 'twin-top' three-speed and overdrive gearbox fitted to their cars in the late 1920s, particularly the Model 629 Limousine. They had also developed a high

A very successful (for Nissan) product of the Nissan/Graham Paige alliance was the Type 70, seen here at a Fifties motor show to demonstrate that Nissan hadn't just come into the motor industry.

quality (but relatively expensive) range of trucks in the early 1930s. The attraction to Nissan was two-fold. First, they were able to add the respectability of a quality American name to their own, adding an international charisma to the Nissan name. Second, they were able to add an existing product line to their range at relatively low cost, for Graham-Paige were looking for revenue at minimal cost in order to save the company. As a result, Nissan bought the designs and the manufacturing plant from their new-found American partner. A most successful product of the Nissan/Graham-Paige alliance was the Type 70, a six-cylinder engined car that was offered in both saloon and open forms.

THE ONSET OF WAR

War came to the Japanese some time before it began in Europe in 1939. Differences with China over several years – including the conquest of southern China and Manchuria in 1931 – erupted into open warfare in 1937. Soon after the start of the conflict, the expansion of all sectors of manufacturing was encouraged by the Japanese government and by September 1937 the Munitions Industry Mobilization Act was passed, along with the Extraordinary Funds Adjustment Act. Their objective was to direct manufacturing industry to support the war effort in China and to aid efficient but under-funded companies in the expansion of their production facilities. Then came the Act for Import and Export of Goods, which controlled the use and movement of manufacturing materials and restricted the production of non-essential commodities.

During 1939, the Japanese Government ordered the nation's industry to concentrate all its efforts on the tools of war and the motor industry was instructed to cease the manufacture of private cars, giving over factory space to allocated war production. This was all part of the Materials Mobilization Plan, which controlled the use of raw materials and manufacturing waste, as well as directing companies to the manufacture of specific products. Nissan were instructed to focus on the manufacture of trucks, as it had only recently bought the know-how and tools from America, but in addition it was given the task of making generator engines, engines for the Navy's motor torpedo boats and aero engines for the Army's training planes. The manufacture of small passenger cars would not re-commence for over a decade.

The Price Control Ordinance came into effect in October 1939, setting the selling price of the products of Japan's motor industry, which was producing vehicles for the movement of goods or for the armed forces themselves. Body sizes of vehicles were standardized and standard production

Produced under the Materials Mobilization Plan, this is a typical example of the trucks produced during the Second World War in the hundreds of thousands by Nissan for the Imperial Army.

costings were established, with a cost calculation committee to regulate manufacturing costs. Vehicles were still supplied to authorized non-government organizations through dealerships, but their profits were regulated by the Ministry for Commerce and Industries to a limit of 10 per cent.

As the Japanese Army advanced through the Philippines, the Malayan Peninsula, the Netherlands East Indies and Burma, Nissan trucks by the thousand accompanied the troops and proved their worth under the most arduous of conditions. Their reliability was discovered by the Allies as they regained territory and trucks fell into their hands. After extensive fire bombing of several industrial Japanese cities, many factories were reduced to rubble or at least had their capacity severely curtailed, though fortunately for Nissan, Yokohama was less badly affected.

POST-WAR REVIVAL

When World War Two ended, Japan had been in military conflict since the early 1930s and the world had seen the rise and fall of the modern Japanese Empire in a decade and a half. In that time, the spread of Japanese equipment and vehicles had been enormous. Now, with the war ended, those same Japanese vehicles, mostly trucks and buses, would be put to more peaceable tasks, such as the transportation of essential supplies and food for the reconstruction of former Japanese conquests, territories and even the islands of Japan itself, for the damage and destruction were widespread and the reconstruction task huge.

The reliability of those Nissan vehicles was now to be tested as never before, for they were bereft of stocks of spare parts and the factories in which they had been made were under the control of the occupying forces. Japan faced its greatest challenge in history, the reconstruction of the nation after the most devastating war ever. But more important to the people of Japan was the reconstruction of the nation's position of greatness in the East and its historic pride. That revival of greatness would now come from the achievement of economic, not military, might.

Until 1952, the Allied Control Commission had overall administrative authority over Japan's political management and the nation was occupied by American military forces. However, the situation was eased considerably for the Japanese by the rising threat of war in Korea, as the American forces used military bases and airfields in Japan as their springboard to Korea as the United Nations expanded their efforts to restrain the ambitions of North Korea.

The primary aim of the Allied Control Commission's management of Japanese industry was much less aimed at control of Japan's economic strength than at reconstruction, and monitoring productive effort to ensure the best use of limited resources. As the population was re-housed and communications were restored, much of that effort employing wartime Nissan trucks, so industrial control was progressively returned to the former owners of manufacturing facilities. Because its factory in Yokohama was undamaged, control of about half Nissan's manufacturing activities was restored to the company in 1952. The remainder stayed under Control Commission management for another three years.

Even under Control Commission administration, the Nissan management was clearly left to do its day-to-day work unimpeded. This involved the renewed manufacture of trucks and setting up a spare parts line for the many existing trucks that had formerly been the property of the Japanese

From Little Acorns

It was from these scenes of utter devastation, after the bombing of Hiroshima and Nagasaki, that Japan had to recover to establish its post-war industry.

Army, but were now being put to use by public authorities, government agencies and private enterprises. The return to truck manufacture began almost immediately the war ended and passenger cars began to be built in 1947, based mostly on pre-war designs.

During the war, the Japanese government had placed all motor vehicle distribution throughout the country into the hands of the Japan Motor Vehicle Distribution Company Limited. This monopoly was dissolved after the conflict but many of the old Nissan motor dealerships switched to Toyota, thinking they might do better with the newer company. This did little to help Nissan's commercial recovery, but the company pulled together in the face of a system that taxed large cars savagely to maximize the limited resources of the nation.

With much of its infrastructure gone, the new Japan was forced to develop an export-led economy, for the foreign exchange earned by exporting whatever it could manufacture would buy foods that it could no longer grow and the raw materials for further manufacture. Nissan's road to success along that path was first impaired by severe labour problems in the early post-war years, which culminated in a 100-day strike in 1953, and the continuing tenure of half its factory by the Allied Control Commission. However, that year was to be the watershed, for new labour relations were established, a new car had been designed and was in production and a new deal was on the horizon.

2 Creating a Sports Car, Creating an Industry

The task of creating a sports car is never easy, even for the volume manufacturers of vehicles who can pour vast sums into development projects while they continue to produce thousands of cars a week from their factories. Imagine then, how daunting the challenge must have been for the engineers of early post-war Japan, coping with foreigners using half their factory floor space, a severely restricted supply line for raw materials and almost no sports car experience to draw on. This was what faced Nissan as they moved towards the recovery of their factory from the Allied Control Commission in the early 1950s.

RISING TO THE CHALLENGE

With the threat of war in Korea, the numbers of Americans based on Japanese military establishments increased and many of them brought British sports cars, such as MG Midgets, into Japan. Nissan's sales and

The MG TC Midget was the archetypal popular sports car upon which Nissan was to model its ideas for the creation of the first post-war Japanese sports car, the CD-3.

Creating a Sports Car, Creating an Industry

engineering teams took note of these little cars and aspired to sell something similar to their military guests, who had no noticeable financial restrictions placed upon their purchases of Japanese goods and who seemed eager to sample the products of their new but temporary environment.

The Japanese camera industry, based on a highly skilled optical manufacturing background, had cashed in on the American presence very successfully, selling Canon and Nikon cameras in particular. These two camera types were based very much on pre-war German designs – the Leica and the Contax. Given that they were copies, they were very good copies and they were much less expensive than the German originals to which they owed their heritage. Better still, the original lenses would fit and couple to the Japanese copies. It was probably the camera industry above all others that persuaded Americans and the rest of the world that the Japanese were capable of fine precision engineering at an economic price.

The Nissan calculation was that if Americans could be persuaded that Nissan could make cars of quality that were solidly designed, reliable vehicles, then the Americans might buy their cars and some might take them home to the United States. Australian sailors, soldiers and airmen were also being drafted into Japan as the threat in Korea grew; these people were even more attractive to Nissan, as they drove right hand drive cars on the left hand side of the road in Australia, just as they did in Japan. Nissan marketing staff calculated, probably correctly, that if the Americans could be persuaded to take a second look at Datsun cars, then the Australians almost certainly would. So the strategy was set: there would be a new Datsun, a sports car designed along European lines.

THE CD-3 STARTS A NEW ERA

Nissan designers first took a long, hard look at what they had already manufactured, then took a look at a wide range of small European open cars and worked out an ideal specification, taking into account Japan's punitive vehicle taxation regulations. The decision was to make a four-seater rather than a two-seater, but the car had to be narrow to cope with the country's city streets. It had to be economical and of a shape that would appeal to sports car enthusiasts. In all aspects, it had to be simple to manufacture, simple to maintain and inexpensive.

The result of this first design study was an outline for a car that was to look a bit like a hybrid between an Italian Fiat roadster of the 1930s and a British Singer roadster of the early post-war era. The radiator grille looked vaguely reminiscent of a cross between that of the Fiat and one from a pre-war Graham-Paige.

Going on to the market in January 1952, this new Datsun was the first Japanese sports car to be built since the end of

The Datsun CD-3, Japan's first sports car. It doesn't look much to us today, but this is the source of one of the most exciting sports cars of our age – the Nissan 300ZX – and all the 'Z' car models in between.

Creating a Sports Car, Creating an Industry

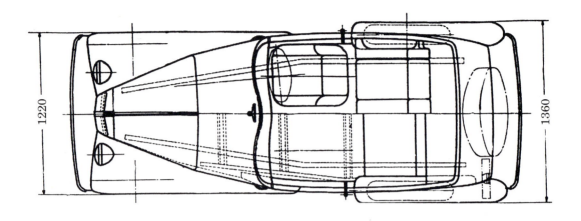

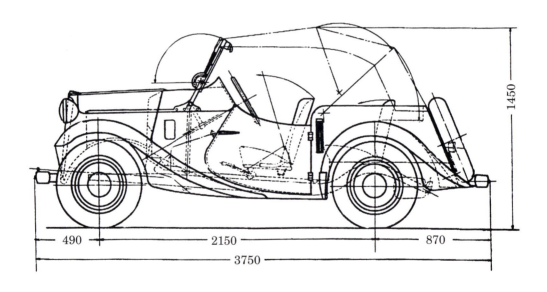

This drawing of the CD-3 shows the simplicity of the car's layout.

hostilities. The project team did not expect to break sales records, but they did expect to generate a little interest from the visiting service personnel and it would certainly provide an object lesson to Nissan in developing their future strategy. The CD-3 would be sold at a price to attract the interest needed for further development.

The chassis of the CD-3 was very simple, the longitudinals being almost straight, except for the arch over the rear axle. In plan view, they tapered inwards to the front end from the front of the front seats and were parallel to the rear. A sub-frame extended forwards to carry the front end of the front springs and the front bumper. Suspension was by half-elliptic springs all round. The front track was 41in (1,048mm), while the rear was 47in (1,180mm) and the wheelbase was 85in (2,150mm). From these dimensions, the reader will observe that this was quite a small car; it was only 124in (3,150mm) long, 54in (1,360mm) wide and 57in (1,450mm) high. These dimensions would allow it to negotiate the streets of most Japanese cities, but it was not the sort of sports car that would match those typical of Europe.

A quite conventional cast iron four-cylinder engine of 60mm bore and 76mm stroke gave fractionally below 860cc displacement. Fuel was delivered by a single carburettor and power output was a puny 20bhp at an unremarkable engine speed of 3,600rpm. The gearbox was a non-synchromesh three-speed unit with reverse, transmitting drive to the rear wheels via an open propeller shaft in orthodox fashion. The gear lever was hardly the short reach quick action type that most European cars equipped with manual gearboxes had – in fact, it was rather a puny bent wire affair with a largish knob on the end that was typical of small low priced saloons on sale in Europe (like the Morris Eight). The handbrake was operated from a short floor-mounted lever and the brakes were, of course, drum all round.

The bodywork was fairly orthodox. The car was of a two door design, with the doors hinged at the rear, very much in the vogue of open car design throughout Europe in the early post-war years. Each door was slightly cut away (there certainly was not much elbow room inside!) and the top line of the doors ran up into the scuttle line, which was shaped with two shallow cowls ahead of each front seat. The windscreen was a fold-flat type with a single wiper mounted at its top edge in front of the driver. Seating consisted of two hinged bucket seats at the front, which lifted up to allow entry into the bench-type rear seat, which was positioned just inside the wheelbase. A simple canvas hood and semaphore direction indicators completed the specification of a car that, to the Japanese at the time, must have seemed quite racy.

THE LONGBRIDGE CONNECTION

As work was progressing on the design and development of the CD-3, the Nissan board realized that an absence from the motor manufacturing business of a decade meant that it was going to be extremely difficult and expensive to produce a competitive vehicle in a market that had advanced so far in design and in manufacturing techniques. So they decided to examine the prospects of finding a manufacturer somewhere else in the world with whom they could strike a deal to borrow designs, as well as perhaps machine tooling and skills. It should be remembered that this was not a new move on the part of Nissan, who had done just this in 1936 with the American Graham-Paige Corporation.

Initial research naturally led Nissan's president Genshichi Asahara to look to the

Creating a Sports Car, Creating an Industry

Austin's Metropolitan, produced to the design requirements of George Mason, President of Nash Motor Corporation. This car gave inspiration to Genshichi Asahara, who contacted Austin at Longbridge and formed a liaison which lasted seven years.

United States, but the only manufacturer there who would at that time (1950–51) have been able to help was the Nash-Kelvinator Corporation, the only company producing a car small enough to be able to run comfortably on Japanese roads. But Nash-Kelvinator were already in negotiation with Britain's Austin Motor Company to design and produce the Metropolitan. George Mason, president of Nash, had decided that America needed a small car, below the position of the Rambler, to meet the needs of the new post-war American market. That car finally appeared in the United States in 1954, after a four-year gestation period.

Next on the list of companies for Asahara to explore was Volkswagen. The German company had produced the prototype of its 'people's car' in 1934, but had now brought it up to date to provide an economical and cheap car for the masses. It was no beauty, but the Volkswagen was already selling in volume in its home market and overseas.

Americans had been amused by the Beetle but many servicemen had bought them and then taken their purchases home. It might have appealed to the Japanese market but its designer, Dr Porsche, was now building his own cars.

So after careful thought, Asahara decided to contact the Austin Motor Company at Longbridge in Birmingham. The need was for a family car that Nissan could first assemble, then build directly from sheet metal. This meant reaching agreement on the provision of members of Austin's workforce to provide local training to Nissan workers at Yokohama, as well as the provision of design drawings, jig tools and presses. Discussions took place in 1952, a year ahead of the formation of the British Motor Corporation. To Austin, the great attraction was long-term financial return for minimal long-term effort. Of course, they would have to provide technical support and place staff in the Yokohama plant to oversee design adaptation, initial assembly training and then the installation of whole manufacturing plant. But after a year or so, Austin would be virtually able to leave the Japanese to it and collect their 2 per cent royalty on sales.

The agreement between Nissan and Austin was signed in December 1952 and announced by Asahara in Japan in 1953. The deal was for the initial assembly, under licence, of the Austin A40 Somerset. An Austin Department was established at Yokohama and several Longbridge technicians and managers were seconded to Japan to train and supervise local personnel in the early stages of the assembly operation, while several personnel from Nissan went to Birmingham to learn about Austin's manufacturing techniques and philosophy. Few of these people had ever seen cars produced in the volumes that left Longbridge, so were in awe of Austin's output.

Creating a Sports Car, Creating an Industry

Cars were initially shipped to Yokohama from Longbridge in kit form. Assembly jigs had been erected at Yokohama under British supervision and in the spring of 1953 the first Nissan-built Austin Somerset rolled off the line at Yokohama. A new era had begun for the Nissan Motor Company as it entered the second half of the 20th Century, manufacturing cars that were contemporary with European models. However, the trade unions that had been encouraged by the mainly American-run Allied Control Commission were not in sympathy with the major reorganization essential to the transition from the old manufacturing ways to the new mass-production techniques of building Austins, and ultimately the 100-day strike took place. It is said that

This is an early A40 Somerset on the line in Japan. The body is being lowered onto the chassis – very carefully, by the look of it!

the Nissan board then induced middle management to form a new trade union; eventually the problem went away and staff went back to work.

The development programme continued and Nissan progressed to the next model in the Austin line. The co-production licensing agreement was for seven years duration, and the next logical step was total manufacture of Austin cars, from the drawing through to the finished product. The next product in the Austin inventory was the new A50 Cambridge, now of unitary construction and of more modern lines. By 1956, the new A50 was in production, though it was August before full production of all aspects of the car was concentrated on Yokohama. That was when the transfer machines were installed for the manufacture of engine cylinder blocks and cylinder heads. These machines would take a raw casting and face it, machine it, bore it, drill it, ream or tap the holes in it and automatically transfer it from one work position to another.

By the time the A50 went into production in Nissan's plant, the workforce had become thoroughly familiar with Austin manufacturing processes, and here A50s are going down the line in quantity.

The most extensive modification made by Nissan to the Austin A50 was the creation of a station wagon or 'van' version, which sold 22,000 examples.

Some modifications were made to the A50 to adapt it to local needs and operating conditions, and Nissan even produced an estate car from their own adaptation of the A50 saloon. While they did not consider it sold very well, it represented about 5 per cent of total Japanese Austin production over the seven year period. They managed to build nearly 22,000 vehicles under the agreement, which came to its end in 1960. There seemed to be no great enthusiasm in the new BMC empire for renewing the deal in 1960 and Nissan had learned a great deal about the mass production of cars, so the two parted company.

In recognition of what they had done during the period of the partnership, the Nissan Motor Company Limited were awarded the annual Edwards Deming Prize for engineering excellence in 1960. Part of the justification for this was the work done in the company on the use of fibre reinforced plastics for the manufacture of car bodies. Glass-fibre was not a new material – cars had been bodied with the material since 1951, as long ago as 1940 the Ford Motor Company under Henry Ford's personal supervision had built a plastic car body using plant fibre and phenol, and glass-fibre reinforced plastics were used in military applications by 1942. Now, Nissan wanted to do it properly and so embarked on a research programme with Professor Atushi Hayashi from the Engineering Department of Tokyo University.

Reasoning that the cost of manufacturing pressed metal panels for car bodies was complicated, time consuming and expensive, Nissan wanted to find a better way. Much of the expense was incurred in making the press tools from which the panels were produced and so, reasoning that multiple curvature moulds would be much simpler to make, they decided there had to be a future in fibre reinforced plastics, provided the strength and durability characteristics were acceptable to the market. As well as researching the strength, elasticity and life potential of the material, Nissan and Hayashi investigated damage and repair techniques, and ultimately produced a serviceable and easily repairable body material. The next sports car they produced would use a body made from this material. In Europe, Lotus was the company credited with being adventurous in the use of fibre reinforced plastics. In Japan, Nissan were to receive much less international recognition for doing its own research and following the same path quite independently.

TO PASTURES NEW

In 1959 Nissan announced a new production model called the Datsun 310 Bluebird. This was a small family saloon, powered by a 1,200cc overhead valve engine that used some of the experience gained with the 1,200cc Austin unit. The line of the Bluebird was distinctly Austin in character, the car bearing more than a passing resemblance to the A50, though on a smaller scale. It was badged as a Datsun and mildly modified to

Creating a Sports Car, Creating an Industry

The Datsun 310 Bluebird was launched for the 1960 season. Based quite closely on Austin design lines, the car was quite well received in North America as being well finished and thought to be a quite respectable small car.

meet American tastes, as that market was now a positive target for the Nissan Motor Company. The new car had Austin-style independent front suspension and a column gear change for the three-speed gearbox. According to *Road and Track* it was a well finished car, offered later as an estate as well as a saloon.

A year later came the replacement for the A50 Cambridge. This was a bigger car than the Austin, designed from the outset to accommodate six people in comfort. It was a three-box saloon, looking vaguely like a Vauxhall Cresta, but without the fins. Powered by a four cylinder 1,883cc engine, with a now respectable 83bhp output, it was also offered with a new six cylinder 1,973cc unit producing 110bhp. A few of these also found their way into the United States and *Car and Driver* reviewed the Cedric, speaking well of its finish and describing it as a genuine six-seater – quite large by North American standards. By 1963, the Cedric was offered with a bigger six cylinder engine of 2.8 litres. This is where the development of the overhead cam L24 engine that was ultimately to power the Datsun 240Z began.

GROWTH AND CONSOLIDATION

Nissan's association with Austin had put it way ahead of its competition at a time when Japan's motor industry was being re-born. Now, it had shown the benefits of that association by introducing the Bluebird and Cedric models and had achieved acceptance of both in its intended major export market. The next step was to achieve growth, either organically or by acquisition or merger. The

In the year following the expiry of the Austin/Nissan licence arrangement, the Cedric was launched, being offered with four and six cylinder engines and well thought of by Car and Driver.

The Prince Skyline and Gloria models were added to the Nissan range with the merger of the two companies. The differences in linguistic expression are demonstrated by this motor show picture of the Gloria, on which the front number plate carries the inscription 'Groria'!

company chose both paths, opening new plants at Oppama in 1962 and Zama in 1965. Then, in 1966, the Prince Motor Company Limited merged with Nissan, bringing its Skyline and Gloria models to the new product range.

Pictures of the Skyline convertible show it to be of a shape reminiscent of the early Ford Mustang, though the radiator grille had a touch of Lancia about it. It was powered by a 73bhp 1,500cc pushrod overhead valve engine that compared well with the British Motor Corporation B Series engine, while the Gloria had a larger 1,850cc four cylinder unit for its larger saloon body. This latter model bore something of a resemblance to the Chevrolet Corvair from the front and was again clearly designed with American sales in mind. The Skyline was certainly imported into the West Coast in some numbers, but the Gloria, even with its optional larger 2.6 litre engine, did not seem to be able to attract North American buyers.

Datsun sports cars were now being sold into the United States too, the SP310 and SP311 making quite decent inroads into the market as the company developed the new Fairlady sports model, the 240Z. As the 240Z was consolidating a position in the American market, things were happening at home about exhaust emissions and vehicle safety. Nissan built an Experimental Safety Vehicle to research into several of these aspects. By 1971 Japan had introduced emission control standards that closely paralleled those proposed by the Clean Air Act in the United States. This political decision helped the Japanese motor industry, as it was producing cars that complied with American emission standards from a very early date.

The two energy crises in the 1970s brought extra benefits to the Japanese motor industry, as the whole world appreciated the need to economize on oil consumption. Datsun cars were notably economical on fuel and as the auto industry in the United States went through trauma and layoff, the rise of the Japanese continued inexorably; in the end American industry sought legislation to control the threat from these imports. Even so, Datsun sports cars continued to sell in the United States and the 240Z was soon acclaimed as the finest mass produced sports car in the world – which is where our story truly begins.

3 The Path to Success and to the Z

In 1958, after a gap of six years since the announcement from Datsun of Japan's first-ever sports car, they decided it was time to try again. After all, car designs had developed apace since their first attempt and the lure of export markets was beginning to tempt all Japanese industry. So work began on designing a new sports car with an all-enveloping body that would carry the looks of the sort of car that appealed to the people in the United States, the prime export target for the new model.

The Nissan Motor Company had begun to export Datsun cars to the United States in 1958, in the wake of Toyota's early export drive. Among the cars offered for sale was the Model 210 Sedan, which had been introduced a year earlier in Japan and was selling quite well, as well as beginning to show its potential by winning the odd sporting event. This model was also being exported to Australia, and in 1958 a 210 won its class in the Australian Mobilgas Trial, a round-the-continent rally. Given the state of the roads in Australia in those days, this victory attests to the ruggedness of the little Japanese car. But it did not sell very well in the United States, as it was considered ponderous and to have soggy suspension, characteristics inherent from its need to cope with the severely pot-holed and narrow roads of its homeland. Even so, Nissan were not deterred and decided that the attractions of the American market had to be worthy of a serious sales drive.

THE RESUMPTION OF DATSUN SPORTS CARS

Nissan's design team began to take a very careful look at which sports cars were selling in the United States, then they studied the design ideas that were coming from Detroit, for the influence of that city on the buying public was considerable and Datsun wanted to sell its products to the widest possible market.

There were really only three contenders at that time which had been converted from the drawing board to metal: the Studebaker Golden Hawk, the Ford Thunderbird and the Chevrolet Corvette from General Motors. The Corvette was far ahead of the other two in market potential, though each one had its dedicated following. But Nissan was after numbers, and so the Corvette and the Austin Healey 100 were the cars on which they were to model their next sporting car.

The new model was to be known as the S211 and it was introduced to the Japanese home market in June 1959. Aimed at the low-priced end of the American sports car market, it had a 988cc engine lifted straight from the 210 and a glass-fibre body, the first ever volume-produced in Japan. The car's line was a sort of hybrid of design ideas culled from the Austin Healey and the Corvette. With a two-tone paint finish, this little car was hardly the most stylish example of automobile engineering,

The Path to Success and to the Z

After careful study of contempory sports car lines, and extensive research into the use of fibre reinforced plastics for automotive bodywork, the S211, with lines drawn from a combination of the Austin Healey 100 and the Chevolet Corvette, appeared in 1959.

but it was a start and a few were sold in the United States.

Encouraged by their first attempt and not put off by its meagre sales, Nissan went back to the drawing board and came up with a development of the 211, to be called the SP212. This car had slightly tidied lines, though its heritage showed through, and a slightly larger engine of 1,189cc with marginally over-square proportions. The bore was 73mm and the stroke 71mm, which gave the potential of rather better power output than was tabled for this model. The actual power output was 48bhp at 4,800rpm on a compression ratio of 7.5:1. This was hardly earth-shattering, but typical of the ultra-reliable small Japanese engines of the time. The cruising speed of most cars in Japan at this time was only around 25mph (40km/h), as the roads were so appalling, so there were some serious lessons to be learned and developed upon. The SP212 was

Next on Nissan's menu of sports car development was the SP212, with tidier lines than its predecessor and improved performance.

The Path to Success and to the Z

the first successful example of a Japanese sports car specifically designed to draw expressly on parts from passenger car models to facilitate mass production at the lowest possible design and production cost, though it retained the glass-fibre body.

THE SP310 MARKS A MILESTONE

The year 1961 was a milestone in the history of Datsun sports cars, as that was the year in which the SP310 was unveiled at the Tokyo Motor Show. It was October and the milestone for Datsun was the huge leap in styling from the 211 and 212 models to this new car. Indeed, it was almost the epitome of the latest in European sports tourers, yet it was built in Japan. The line was remarkably similar to that of an internationally famous car introduced at the London Motor Show a year later: the MGB. This new Datsun drew applause from the international press and sports car world and was to be the car that really set the foundation for the development of the 240Z in just under a decade.

Similar progress had been made by Nissan in the area of engine development. The rather feeble 48bhp of the 988cc engine in the 210 was gone. Now, the new 1,488cc power unit, with a bore of 80mm and a stroke of 73.7mm, was tuned to give a quite respectable 85bhp on a 9:1 compression ratio running at 5,600rpm. That compared very favourably with the later MGB's 95bhp from its 1,800cc engine. The four-speed gearbox and this engine, together with a notably improved suspension, made the SP310 quite an exciting little three-seater. Three-seater? Yes, it had a transversely positioned third seat behind the driver and front passenger until 1963, when the third seat was abandoned and the car became an orthodox two-seater.

The styling of this new car was largely based on a series of intensive studies carried out by the Nissan design team to make it attractive in the eye of its mainly American market. One Nissan executive is said to have commented that his company was told: 'Send us something enchanting' so they went to work on Project SP310. While it had to appeal to the home market as well as to its export markets, this new car needed to follow internationally accepted conventions of styling which, in the world of sports cars, had a very strong European influence. The Chevrolet Corvette was the only all-American sports car to enjoy true international success.

On the road, the SP310 made quite an impression and was as far ahead of its

The SP310 brought the Datsun range right up to date in style and line, drawing praise from the international press and setting Nissan on the road to the 240Z.

The Path to Success and to the Z

Proving the SP310, known in the USA as the 1500 Sports, came from its use and proven durability in rallies.

predecessor in handling and performance as it was in styling. You would barely have believed that the two cars could have left the same factory in succession. Offered for sale in the United States as the Datsun 1500 Sports, this car really was the catalyst of Datsun's programme to produce an international sports car. But the Japanese manufacturer knew that market progress would not be fast and so put a positive development programme in place to create that world sports car. In the meantime, the SP310 sold a total of almost 7,000 before it was replaced.

The creation of the SP310 was in many ways a landmark for the Datsun name. It was the first Datsun to come anywhere near meeting the performance criteria essential to securing sales in the United States, the first to look the part and the second to sell in anything like reasonable numbers. When they first launched into the American market, the target was to sell 5,000 cars of all types in the first year, but enough of the sporting S211 to make a mark in that market. That first attempt was a dismal failure, partly because the Japanese had miscalculated the market demands and partly because, as a result, they had not come up with acceptable specifications that would compete. In addition to that, there was the lingering perception that the Japanese were good at copying but no good at innovation, and unreasonable questions about how they actually made them. This last point arose because the Japanese had proved themselves tremendously resourceful in the early post-war years to build up a manufacturing industry. Notwithstanding the benefit of huge international re-construction subsidies in those early post-war years, the subsequent results speak for themselves. In order to bolster their prospects, Datsun USA was formed to launch the SP310 and their success with this model, albeit modest, is a matter of history.

THE SP311 HERALDS THE NEW AGE

At first glance there seems to be little difference between the SP310 and its successor the SP311. But the board of Nissan were now getting serious. They had seen that the 310 had made a mark on the market and had proved itself a fairly respectable performer, measured against other Japanese imports and even against some of the European products entering North America. It had also been demonstrated that a sports car could be built while drawing on the passenger car parts inventory, provided the chassis design was right. Now it was time to build on that very modest start. And now it was time to bring in Albrecht Goertz.

Before the SP311 was unleashed upon the world, a new Datsun coupé was announced at the Tokyo Motor Show in

The Path to Success and to the Z

COUNT ALBRECHT GOERTZ – STYLIST OF THE DATSUN 240Z

Albrecht Goertz was born into the German nobility at Brunkunsen, in Germany, the second son of the family. Born in 1914, he had a clear awareness of what was beginning to happen in 1933 as the result of Adolf Hitler's rise to power. Goertz decided that he would be much better off taking his chances in the New World, so he emigrated to the United States.

Deciding to study design, Albrecht Goertz went to work at the automotive design studio of Raymond Loewy. From the late 1940s he worked with Loewy on the lines of the 1950 Studebaker range, going through to the 1953 model year cars. Setting up on his own, his first major design of note was for a car maker in his homeland. The car he created – the BMW 507 sports two-seater – came to be noted as one of the milestones of automotive design in the 1950s and is today revered as a true classic.

Going on to work with another great name in sports cars – Porsche – he finally decided to return to the United States. However, his stay was short-lived, for his reputation had reached Japan's shores and the board of Nissan Motor Company Limited invited him to design a car for them. It was to be a coupé version of the SP311 sports car, to be known as the CSP311. The car was announced to the world at the 1964 Tokyo Motor Show and went on sale in March 1965, preceding the SP311. Hardly surprisingly, these two new models both featured Porsche synchromesh in their new four-speed gearbox. The CSP was not the international seller that even the later four-cylinder Datsuns were, but the best was yet to come.

After a brief sojourn at Toyota, designing the 2000GT, one of the most delightful car outlines ever to emerge from Japan, Goertz returned to Nissan, with the charge of designing another car, with a completely free hand. His choice was a sports car, a two-seater, but a car with remarkably close dimensions to one of his former employer's products, the Porsche 911. That new car was to achieve fame as a fine car and as the largest volume mass produced sports car ever. Albrecht Goertz had established himself as part of a legend in sports car history.

This picture of Albrecht Goertz (on the left) was taken in recent years and was provided by 'Z' car enthusiast Len Welch, whose car it is.

The Path to Success and to the Z

September 1964. It was called the CSP311 and was a striking machine that was intended to turn heads and show the world that Datsun sports cars were a force to be reckoned with. The styling of this new car was largely the work of Albrecht Goertz, a German count whose industrial design house was responsible for the BMW 503 and 507 models, then later the Toyota 2000GT. His name would come into contact with Datsun again before long, as he was brought in to advise later on the design of the 240Z. Goertz's work on the Toyota was hailed by the world's motoring press and by the industry itself as close to genius. It certainly was an outstanding car, and if Toyota had been serious about production of the car for volume sale, it could have given the Datsun 240Z a serious run for its money.

Albrecht Goertz was given a free hand in designing and developing the CSP311, for Nissan were after real sales and were even ready to pitch examples of their product into competition. Goertz persuaded his employers to embark on a hitherto untried design process, which had been used in Europe and North America for years, but not in Japan. They built a full-size clay mock-up, to allow views to be taken of a car in as near a 'real' situation as possible.

This approach also allowed the impact of changes to be viewed in a less artificial and conceptual frame of mind. It provided for a more assured design approach, with less risk of expensive mistakes. Out of this exercise came the next production model in the Datsun line, the SP311, a car not far removed from the styling of the 310, but one that laid the ground for the introduction of what many have described as the most successful mass produced sports car ever. It was announced in May 1965.

The SP311 differed from its predecessor in several ways, mostly under the skin. The engine of the new model, developed directly from the unit in the CSP311, was already proven and reliable. It was a four-cylinder pushrod of 1,595cc, fuelled by a pair of Hitachi carburettors that helped the 9:1 compression ratio to generate a healthy 96bhp. This was 1bhp more than the original MGB, which was now on the streets of the United States and with which this new Datsun would inevitably be compared.

Some people have suggested that the design of the SP311 was cribbed from the

Out of Albrecht Goertz's CSP311 came the SP311 Sports: with its 1600cc engine and 96 brake horsepower, it was a car to challenge the MGB head on.

33

MGB; some even thought it a near-clone of the British car. But nothing could be further from the truth, as the car's immediate predecessor, the SP310, upon which the styling of this new model was clearly and closely based, was designed and released in Japan before the MGB emerged from Abingdon. Work on the Datsun SP310 probably began before the Abingdon team started development of the MGB, not least because the 310 was on the market a full year ahead of the British car.

The new car was said in some quarters to offer narrow seating, yet that in the MGB was little wider and its predecessor, the MGA, offered much the same level of seating. The SP311 was only marginally narrower than the MGB and that was probably a sub-conscious act of design, taking account of the narrow streets of many Japanese cities. The styling of the car was certainly in keeping with its age, though, and under the skin, it was right up there with any of its adversaries. It had a ladder-type sub-frame, torsion bar front suspension, Porsche synchromesh on all four gears, a diaphragm clutch and lubrication-free drive shaft components. In the words of one motoring journalist, the SP311 'oozed reliability, not oil', for one of the problems with the MGB was its inability to retain its lubricant.

THE SP311 ON ROAD AND TRACK

In its first year on the market the SP311 sold nearly 5,500 units, set against 7,000 of the 310 in two and a half years, so positive

The Alfa Romeo Giulia Spyder (below) was the middle-market challenger for the SP311, whilst the Triumph TR4 (above) was a head-on contender, and the Datsun began to break ground to establish quite a respectable market for itself.

progress was being made. Badged in the United States and in Europe as the Datsun 1600 (very few in fact found their way into Europe), the new model was a head-on challenge to the MGB, Triumph TR4 and Alfa Romeo Giulia Spyder 1600. The Alfa Romeo was just about twice the price in America, so was almost immediately discounted, but the price of the MGB and the Triumph made them positive adversaries for the Datsun 1600 – quite apart from a broad similarity in styling.

The Datsun had wind-up windows, a heater, a tonneau cover and a radio. What was more, it was a 100mph (160km/h) sports car with the roadholding and stopping power to match. With disc front brakes and heavily finned rear drums, it was capable of stopping in less than 30ft (9m) and in a straight line from 30mph (50km/h). During its second year of sale, the 1600 sold over 6,250 examples and as 1967 approached the car's popularity grew, with the result that there was to be another development in the line, leading to the SR311 2000 model.

The reliability of the 1600 soon earned it the opportunity of try-outs in racing at club level. The pure sports car fraternity of California had taken its time to appreciate the potential of the Japanese sports cars, but the fact that the 1600 ran so well and did not need constant oil top-ups as the result of leaks where gaskets should seal meant that they saw it as 'worth a go'. Datsun USA gave support to all sporting activities involving their cars, including racing and rallying, and it was not all that long before the spectators were watching Datsuns lead MG, Alfa Romeo and even Porsche cars round the track.

The performance record of the 1600 was growing every weekend, as racing success followed racing success, and rally victories notched up, too. As the competition success list grew, so more individuals began to take a second look at the Datsun SP311, then they went out and bought one! It is hardly surprising that they did, when the price tag for a 1600 on the road in 1967, just before the 2000 joined it, was a mere $2,546. When you consider that the MGB and the Triumph TR4 were both over $3,000 at the time, the Datsun 1600 suddenly became a very attractive proposition. The market reaction was such that a new model was added to the stable in 1967.

THE NEXT LINK IN THE CHAIN

The final link in the chain of succession from the S211 to the 240Z was added to the Datsun product range in March 1967. This was the SR311 and it brought with it detailed improvements to the specification and modifications that were designed to meet the United States Federal Motor Vehicle Safety Standards. These included new door releases, a larger safety-glass windscreen, seat head restraints, new mirror positions and a collapsible steering column. The SP311 and the SR311 were now offered side-by-side as a two-model sports car range, the SP in the 1,600cc class and the SR a 2 litre.

Carrying with it all the design features of the 1600, the new 2000 roadster was an even better match for both the MGB and TR4, costing a mere $330 more than the 1600. The new car enjoyed a twin overhead cam power unit of 1,982cc, with a bore and stroke of 87mm x 82.8mm and a compression ratio of 9.5:1. The power output of this new motor was an amazing 135bhp. It is not therefore surprising that it began to eclipse the MGA and the TR4. It was lighter, too, weighing in at only 2,006lb (910kg). The twin Solex carburettor option added another 15bhp, bringing the maximum output to 150bhp and giving the Datsun 2000 at least

The Path to Success and to the Z

The SR311 was the last of a line, basically an updated and enlarged engined SP311, which brought with it detail improvements to the vehicle's safety specification to meet new US standards. It proved to be a fine car and popular with American sports car enthusiasts.

Interior of the above car.

The Path to Success and to the Z

A Hardtop version the SR311, the 2000 Sports, was offered in 1968, to stay ahead of the anti-open-car lobby.

comparable performance with the Alfa Romeo Spyder. This new version of the Datsun sports car consolidated the very keen value-for-money philosophy of the growing Japanese car industry.

As the 2000 established its place in the market, so the SP311 was updated in body and chassis specification. Essentially, the same vehicle type went down the line, the only difference when it came off the line being whether it had the 1600 pushrod engine and four-speed gearbox or the 2000 engine and five-speed gearbox. Thereafter, the relevant badges were fitted and the individual car became either a 1600 or a 2000.

One last change to the SP311 and SR311 models took place in November 1968, shortly after the Tokyo Motor Show. Both were offered in hardtop form as well as roadsters. The age of the coupé was coming, and in the United States market it was being helped along by such people as Ralph Nader, who set himself up as a one-man campaign against the evils of the motor industry in America. His voice, and others like it, was heard and the American legislature brought about far-reaching changes to motor vehicles, some necessary and some not. Among the less necessary, as history now tells us since the revival of the drop-head, was the pressure to ban open cars. Seeing the problem ahead, and the possible damage it could do to their hard-won market, Nissan answered with the hardtop version of their sports cars. But by now, something much more exciting was in the wings and the Datsun name would soon take centre stage.

THE RETURN OF ALBRECHT GOERTZ AND AN ADVERSARY THAT WASN'T

Albrecht Goertz had been retained to create the Datsun CSP311 in 1964. Such was the market reaction to that model, and the confidence it gave the manufacturer that the path they were following was the right one, that Goertz was recalled to style the next sports car in the Datsun line, aimed at

The Path to Success and to the Z

becoming the sports car of the 1970s. The decision to recall him was based as much upon the success he had already achieved with the CSP311 as on the recent outstanding piece of design work he had completed for Datsun's main rival, Toyota, in their 2000GT. The Toyota 2000GT deserves space in this book simply because it came into being and could possibly have killed the 240Z stone dead in its tracks. Even now, it is amazing that Toyota did not seem to see that they had a world-class winner on their hands, yet they chose not to manufacture it in quantity. The car was designed by the studio of Albrecht Goertz.

It was in 1964 that Toyota saw the case for developing a real sports car. They had watched the progress of arch rival Datsun and the expansion of their share of the sports car market from a damp start to quite a respectable level of sales in America, where there was an inherent resistance to overcome, quite apart from any natural apprehension that greets a new and unknown product. Toyota wanted part of that market share and a sure way of making progress in that direction was, in their eyes, to hire the man that had a design influence on Datsun's success. That man was Albrecht Goertz.

Bearing in mind that there were affluent Japanese who wanted a home produced sports car of respectable style and performance, quite apart from the international market that lay out there, Toyota's choice of engine size was 2 litres. Car tax in Japan rose to astronomical levels over that engine size and the 2 litre sports car was internationally popular, taking into account the current products of Alfa Romeo, MG and Porsche, for example. It appears that some of Goertz's design ideas were built up into a prototype for Nissan by Yamaha. Before ensuring security of copyright, Nissan rejected the Yamaha prototype and proceeded down the route of the SP/SR311.

The Toyota 2000GT, Albrecht Goertz's masterpiece for Nissan's prime competitor. For reasons unkown to most of the world, it never went into series production and so gave back the initiative to Nissan.

The Path to Success and to the Z

The Yamaha-designed Toyota 2000GT engine.

The consequence of that was that Yamaha made an approach to Toyota and sold them the prototype. Goertz was then called in to develop it into what was to become the Toyota 2000GT.

The development of the 2000GT was a co-operative affair with Yamaha and Goertz from the beginning. Goertz handled the styling, Toyota handled the chassis design and Yamaha produced the engine. The chassis was modelled on the Lotus and Triumph principle of a single longitudinal member – a box section backbone. Lotus used them to support their glass-fibre bodies, while Triumph used the concept as a space provider in such models as the Herald saloon and the Vitesse four-seat drop-head. Extending out from the backbone, in similar fashion to the chassis of the Lotus Elan, were the supports for the suspension, with unequal length A arms linking coil springs and shock absorbers and quite heavy anti-roll bars at all four corners. This gave a firm, positive ride with handling pretty close to that of the Lotus.

The Yamaha-designed engine was an in-line water-cooled six of 2 litres capacity, with twin overhead camshafts. Equipped with cross-flow porting (the inlet on one side and the exhaust on the other), the engine had three twin-choke Solex 40PHH carburettors, made by Mikuni under licence, and two three-branch exhaust manifolds. In standard form, the engine produced a very impressive 150bhp with a 7,000rpm red line, though it was clearly capable of being tuned to a level significantly beyond that. A close-ratio five-speed gearbox transmitted that power to the 4.375:1 final drive and endowed the car with a 0–60mph time of under 9 seconds.

Some have said that the 2000GT was a virtual crib of the Jaguar E-type, but that is too simplistic an approach. As you look closely, you can see traces of the line of the Ferrari Daytona coupé (which came somewhat later) in the rear quarter. The front end might have a trace of E-type about it, but the car fortunately lacks that 'broken-back' look of the E-type's profile, where the front wing line abuts the accentuated beginning of the line over the rear wheels. Bereft of excessive chrome trim generally, the 2000GT did suffer from an oddly shaped radiator grille with over-chromed headlamp trims, but otherwise it was an extremely pretty car. The magnesium alloy road wheels further enhanced the looks of this very well proportioned hatchback coupé.

Priced at nearly $7,000 – more than you would pay for an E-type or even a Porsche of the day – the car was also only offered in right-hand drive form, so very quickly the American market began to ask if Toyota were serious about marketing this new model when it launched it in 1967.

This slightly overhead view of both the Jaguar E-Type (above) and the Toyota 2000GT (below) show some similarity between the two cars, but Goertz's design was distinctly his own.

The Path to Success and to the Z

But Carroll Shelby, a name already linked with the successes of Aston Martin and who was to achieve huge success with the Ford Mustang and the immortal Cobra, saw the potential for this Japanese coupé and went to work preparing it for racing, extracting over 200bhp from the engine. The car also secured racing successes in Japan and a 10,000-mile speed record at over 128mph (206km/h). Yet Toyota built only 337 examples. Why it did not go into volume-production we shall never know, for it had the publicity and it had the specification.

Two examples of the 2000GT were built as roadsters for the James Bond film *You Only Live Twice*, much of which was filmed on location in Japan. That the car was only built in right-hand drive form must have had an adverse effect on its sales potential in North America and Continental Europe, but Great Britain and Australasia could have been very good markets, yet it was not pushed in either location. The Toyota 2000GT could so easily have been: 'The sports car of the 1970s' as the consequence of a copyright slip, but because they either ignored the potential, could not sort out the deal with Yamaha, or did not realize what they had, Toyota let the initiative go back to Datsun – and the Nissan Motor Company were not about to make the same mistake twice!

In late 1967, Albrecht Goertz was called back to Tokyo to go to work on the next generation of Datsun sports cars. The success of his work can be measured by the fact that while his Toyota 2000GT was certainly the first true Gran Turismo built in Japan, the Datsun 240Z was to become known as the first truly successful mass-produced sports car in the world. What an accolade! And how different history could have been!

4 The Z Age Begins: the 240Z and the Fairlady Z

The SP and SR 300 Series sports cars had served Datsun well and had been popular sellers at home and in North America, but by the late 1960s it was apparent that these two models were falling short of the demands of an ever more discerning market that called for ever higher performance standards. So, Nissan's designers, under the guidance of Albrecht Goertz, went to work to create an entirely new, faster, safer and more attractive car than they had ever produced so far. November 1969 saw the birth of a new age for Nissan and a whole new sports car market worldwide, with styling based very much on the latest from Europe, particularly Ferrari, and engineering relying more on computer aided technology than any Japanese car until now. Taking the automotive world by storm, here was a sports car for the 1970s – the Datsun 240Z.

CREATING THE CONCEPT

The new car was seen by its creators as a product for both home and export markets, though they may not have been quite aware of the huge success potential of their new creation. In the home market, it was to be called the Fairlady Z and was to embody a number of specification variations from the ultimate export variant, the 240Z, which itself would be offered with a very wide range of options to suit the many markets in which it would be sold. Basically, there was to be a 2-litre model available at home and a 2.4-litre version for export to North America, Australasia and Europe.

From the beginning of the preparation of a concept for the new sporting model, it was clear to Nissan's management, design and marketing teams that export would be the key priority, as the vast majority of the car's predecessors had been exported, most of them to the United States. And, of course, foreign currency earnings were of paramount importance to Japan's continued economic growth. In the wake of the demise of the Austin Healey 3000 and its erstwhile market successor the MGC, the Japanese team saw a growing opportunity for a market in Europe. After all, the products of Mercedes-Benz, BMW, Alfa Romeo and British Leyland were all likely to be more expensive than a new Datsun, the performance of which would be up to anything it had to compete with.

An intensive demographic study of the potential markets for Japan's newest sports car was carried out before anyone went near a drawing board. Research included investigations into the preferences of potential buyers for open or closed sports cars, two- or four-seat accommodation, interior fit requirements, income brackets of potential buyers and the current lifestyles of those people expected to be most likely to buy. Then, when the outline of the new model was being considered, the question of what size engine it would require arose, taking into account

The Z Age Begins: the 240Z and the Fairlady Z

the performance needed and the models it was most likely to succeed in the various markets at which it would be aimed. North America and Europe were clearly early target markets and so the styling had to carry something of the line of a desirable sporting model recognized in those markets.

CREATING THE CAR

One of the more successful sports car outlines of the 1960s was that of the Ferrari 250GTO of 1963. Its flowing lines were the envy of many sports car makers and were widely emulated. So it was hardly surprising that Albrecht Goertz, when looking for a classic line as a basis upon which to style this new model, chose the 250GTO as the starting point. Of course, the functional requirements of the new project HS30 would be substantially different from those of the 250GTO, the Datsun being a road-going sports car that was required to convey its occupants with comfort and speed whereas the Ferrari's sole objective was to cover the ground at the highest possible speed, with little regard for the comfort of its normally sole occupant, whose prime task was to win a race.

Just two examples of the many variants of pre-production designs considered before the final version of the 240Z was arrived at.

The Z Age Begins: the 240Z and the Fairlady Z

The now-demised Austin Healey 3000 (above) was the prime target for the new Datsun 240Z, though Nissan's design engineers had the Alfa Romeo GTV (below) in their sights, too, as that car's performance envelope was not so far removed from that intended for the 'Z' car, though it was, of course, a 2+2.

The Z Age Begins: the 240Z and the Fairlady Z

The Ferrari 250GTO, despite being six years old by the time the 240Z appeared, was a classic design and gave much to the line of the new 'Z' car.

Compare the lines of Datsun's new 'Z' car with those of the 250GTO (left) and the relationship of line becomes apparent, though Datsun, and Albrecht Goertz, did it well and so the car does not come across as an immediate and outright 'crib'.

The features that had to be allowed for included a raised interior roof line, to provide a sliding seat rack under an upholstered seat and to accommodate a range of driver heights, the ground clearance essential to a roadgoing production car as distinct from a sports-racer, as well as allowing for a production car's exhaust system with silencer. Then there was the provision for silencing and trim materials, as well as deeper section roadgoing tyres. The low nose line of the Ferrari could not be duplicated, even if Nissan had wanted to, as the cooling system of a car that would be driven in traffic was significantly different from that of the 250GTO and the positioning of headlights had to meet the legal height requirements of a wide range of countries.

Despite all these operational and technical requirement differences, the car that emerged as Project HS30 looked remarkably similar to the vehicle that inspired it and the years have confirmed its classic lines. As with any car design, several styling exercises were examined before the final definitive shape was settled upon, including the processes of wind-tunnel testing and computer-aided line refinement. Having once established the line, the computers were now tasked with aiding the structure of the new model, which was to be a steel monocoque two-door coupé with a hatchback tailgate.

MEETING THE MARKET

After a long and exhaustive series of investigations, probably the most expensive and detailed of any Japanese motor company thus far, the decision was made to create a vehicle that would not offend the growing American lobby against the open car and yet would seek to fill the market void left by the demise of the Austin Healey 3000, which had not been successfully filled by the MGC, a hybrid of the MGB body shell and the Austin Healey engine. Despite both the Austin Healey and the MG being assembled at Abingdon, the handling characteristics and performance of the MG did not come anywhere near those of the car it succeeded. But then, it could never really be expected to match the car it was intended to replace, as the MGB had been designed as a 1.8-litre car and the 3-litre engine from the Austin Healey upset the weight distribution of the car into which it was installed, to say nothing of the effect on the suspension.

Fuel consumption was an item constantly borne in mind by the design team working on HS30, with the result that they concluded they could achieve the performance of the Austin Healey 3000 without going to the full 3-litre engine size. This, of course, was partly due to the fact that the HS30 was a coupé, thereby having a lower drag coefficient, and partly due simply to the advances made in engine design since the creation of the C Series six-cylinder BMC power unit (upon which the Austin Healey 100/6 and 3000 engine was based) in the 1950s.

Because it was decided to make a version of the new model for the home market as well as for export, two engine installations had to be allowed for in the design. The export variant was to be known as the 240Z, fitted with the 2,399cc Model L24 single overhead cam power unit, while the home market Fairlady Z was to be powered by the L20 single overhead cam 2-litre six. In external dimensions the two engines were generally similar, so the installation was no great problem. And it was not long before a more sporting version of the home market car was developed, to be known as the Z432, powered by a twin cam 1,988cc unit and featuring a lightweight body, which was later extended into the Z432-R, the ultimate competition variant of the 240Z developed for international rallying and racing.

Meeting the market for the 240Z meant that Datsun had to face up to the fact that potential customers in the new car's major target market – the United States – had for many years been accustomed to buying European sports cars. Datsun had made inroads into that market with the SP and

The 1800cc MGB Coupe was the nearest thing that Britain could throw up against the 240Z as the Japanese car took the sports car market by storm.

SR series models, but those two were both broadly based on British cars and were able to cash in on the sporadic availability of Austin Healeys, MGs and Triumphs. They enjoyed a price advantage as well, but many traditional enthusiasts were keen to dismiss the Japanese models as advantage-taking cribs of sports cars. The 240Z was to change all that.

The European vehicles that the 240Z had to meet head-on included some that were not true adversaries and some that were. It had to be priced attractively enough to persuade sports car fans to divert their attention from such machines as the Mercedes-Benz 280SL coupé at the most expensive end (the perfect 'clone' sporty car), the Porsche 911T for true performance (but still expensive) and the Alfa Romeo GTV. They also wanted to compete with the lower performance and much lower price of the Fiat 124 Coupé, Opel GT, MGB GT and Triumph GT6, which cost between $3,000 and $3,500.

THE 240Z IN THE METAL

Picking up ideas from the best of Italian styling, together with a few ideas from German and British engineering, Nissan's styling and engineering teams came up with a fine steel-bodied sports car that fitted dimensionally right in the middle of the pack. The wheelbase of the 240Z was 0.3in (8.5mm) longer than the MGB and 1.5in (38mm) shorter than the Porsche 911. It was taller than the MG, but lower than the Porsche and wider than both. It looked strikingly like a slightly scaled down Ferrari 250GTO. In appearance it was, in its time, a stunner – and this new car was no sheep in wolf's clothing!

The 240Z's rear suspension took up a lot of space, but proved to be solid, reliable and effective.

Two features of other Datsun models were brought to the new 240Z. The first was the engine from the 510 saloon and the second was the front suspension from the 1800 Laurel saloon. Neither was installed as it stood, of course, but the value of making parts interchangeable – and thus manufacturing cost effectiveness – came from the engine, as the 1.6-litre four-cylinder acquired an additional pair of cylinders to become that very effective 2.4-litre overhead cam six in the form of new L24 power unit. Similar benefits were drawn from the Laurel suspension, though quite clearly it was beefed up to meet the more stringent requirements of this new sports coupé. The MacPherson strut suspension design had by now proved itself in many cars around the world and had established a reputation for being easy to maintain and inexpensive to manufacture. One other 'borrowed' component was the gearbox – while very much based on the four-speed all-synchromesh unit of the 510 model it was essentially a design upgrade, with much stronger internals to accommodate the higher torque output and a different set of ratios to match the performance requirements.

While the styling of the 240Z was distinctly Ferrari, this new car was certainly no clone. For starters, it had a unitary steel body and chassis unit with enough strength to mount a large, wide tailgate, something the lightweight Scaglietti-bodied 250GTO with its tubular chassis could never have supported. The floorpan was quite conventional, with longitudinal box sections lent rigidity at the front by the very strong front bulkhead structure. At the rear, the Chapman-type struts were mounted in pressed steel towers that ran into the rear wheel arches. An upright transverse steel panel was positioned behind the two seats to give rigidity at the rear. The upper section of the rear body had to support the large hinged tailgate, while the lower underbody section, via a very clever series of lightweight sheet steel fabrications aimed at keeping down the production cost, supported the rear suspension and final drive assembly.

This solid, chunky design, using remarkably lightweight and inexpensive design features, allowed the 240Z to be priced at $3,600 in the United States, putting it into distinct contention with the British end of the American sports car market, vying for sales with the smaller engined and thus much lower powered MGB and Triumph GT6. Because the market research and product development had been so carefully conducted, it should be no surprise to learn that the 240Z would win hands down.

RIDE AND COMFORT

Whenever any vehicle manufacturer sets out to design and build a sports car, the question of defining what a sports car is inevitably arises – and the answer is always different, because each designer has a different view and a different set of priorities. In addition to that, the demands of the various markets at which that car is to be aimed must also have an influence. The key parameter for the Datsun 240Z was that it should sell well in Datsun's biggest export market, the United States, so it needed the handling characteristics of a good European sports car and the recognized American market creature comforts to persuade would-be buyers to take a second look.

The choice of MacPherson strut front suspension was a very logical move for what was hoped to become an international sports car. It would cope with a wide range of road surfaces and still provide the driver with good steering response. Installation time on the production line was minimal: the bought-in component cost was low and, as has already been said, maintenance time and cost were low, too. The principle of its operation is that the spindle at the bottom end of the strut is attached to a ball joint and connected with a lower link to the suspension member through a rubber bushing. Forward and rearward motion of the transverse link are restricted by means of a compression rod fitted to the chassis.

An anti-roll bar completed the front suspension specification. The movement of the front end was controlled, yet because it was relatively soft in comparison with many sports cars, it catered well for the demands of the more comfort-conscious American buyers. Many of the 240Z owners in this market would buy the car not because they wanted the handling characteristics of a high-performance sports car but because they wanted its looks.

The rear suspension is described in some quarters as 'Chapman struts' and in others as 'MacPherson'. Like the front, there was

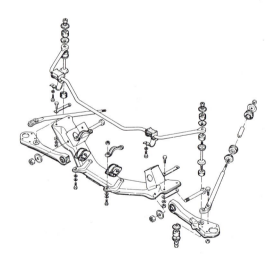

The MacPherson strut front suspension was already proven as a design and would be cost-effective in manufacture, so was an obvious choice for the new 'Z' car.

The Z Age Begins: the 240Z and the Fairlady Z

the combination of a shock absorber running up the centre of a coil spring, with a capping plate at the top and a centre bolt at the bottom which made up a suspension strut to which was attached a wishbone link at one end and a rigid mounting into the bodywork at the other. The general trend has been to describe such a set-up at the rear as 'Chapman' strut because the late Colin Chapman, of Lotus, devised that concept for use on his own sports and racing cars, as well as for the Vanwall Grand Prix car of the 1950s.

On the 240Z, the bottom wishbone articulated with the hub carrier, which was attached to the bottom end of the suspension strut, while the top end of the strut was bolted in to the top plate of the tower. The differential was carried on a mounting plate at one end and a transverse leaf spring at the other. Again, an anti-roll bar took control of transverse movement. The consequent ride of the 240Z was generally very good, though early reports came back that it was a little too soft, even for the semi-sporting end of the American market. However, those dedicated to real sports cars were not quite as impressed as Datsun had hoped, but they did have the good grace to conclude that the problem was not insoluble.

The single overhead cam engine, with a bore of 83mm and stroke of 73.7mm, a compression ratio of 9.0:1 and a torque rating of 145.7lb/ft at a quite modest 4,400rpm, was certainly not over-stretched. Fuel was delivered by a pair of Hitachi-made SU carburettors, produced under licence. Under a heavy foot, its 150bhp soon tested the ride characteristics of the suspension and showed that the nose might lift a little more than the driver anticipated. With factory standard wheel and tyre equipment, the tendency to understeer under aggressive cornering was more than might be expected, though it was still controllable, unless you took a corner on near full throttle, when it was inclined to ignore your attempts to manoeuvre via the very precise rack-and-pinion steering. As long as you were aware of the potential consequences, lifting the foot would convert gross understeer into a sharp state of oversteer – fine if you were ready for it but potentially fatal if you were blissfully unaware.

Despite the handling limitations of the first 240Z, it still appealed to most would-be buyers in the United States, simply because of its styling, its straight-line performance and the sheer satisfaction to be derived from driving a really sporty looking car that was in the same price bracket as the MGB, but that would leave the MGB standing in almost any situation. In fact, here was a car that would leave any of the lower priced British sports cars standing, though a five-speed gearbox as a standard feature would have been a distinct advantage (this was soon offered as an option, but was not part

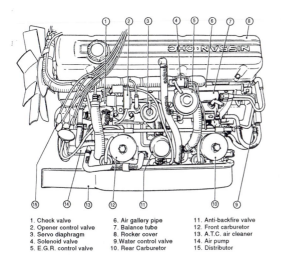

1. Check valve
2. Opener control valve
3. Servo diaphragm
4. Solenoid valve
5. E.G.R. control valve
6. Air gallery pipe
7. Balance tube
8. Rocker cover
9. Water control valve
10. Rear Carburetor
11. Anti-backfire valve
12. Front carburetor
13. A.T.C. air cleaner
14. Air pump
15. Distributor

The relatively simple single overhead camshaft engine of the 240Z, with its SU-type carburettors, was to prove reliable, gutsy and nearly unburstable.

The Z Age Begins: the 240Z and the Fairlady Z

Aston Martin's DB6 (above) and Jaguar's E-Type (below) were really the only two British sports cars that, ignoring price, would be capable of outrunning the new Datsun.

of the initial specification). Even so, it was going to take an Aston Martin DB6 or an E-type Jaguar to out-pace and out-perform it.

To complete the performance specification of the 240Z, and recognizing that you can only ever drive a car as fast as you can stop it, the front wheels were brought to a halt by means of a pair of Girling 10.7in (272mm) disc brakes made under licence in Japan by Sumitomo – who later bought out the mighty Dunlop Tyre Company. Slightly surprisingly, except for the consideration of production costing, the rear brakes were 9in (229mm) drums. Despite the car not having an all-disc brake specification, the stopping power of this machine was pretty astonishing, with a braking distance of only 27.5ft (8.4m) from 30mph (48km/h). From 80mph (130km/h) the overall stopping distance was 287ft (87m) – and in a straight line.

INSIDE THE Z

There are two aspects of any car that help to sell it before the potential purchaser looks at the performance. That is especially true if the real performance is either not that important, or if it is not particularly easily usable in the environment in which the car is to be used anyway. The American market was generally just such a market, where many cars would be bought on looks and interior trim, while a much smaller number would be bought for out-and-out performance. Datsun's target market was the United States and that philosophy was one of which they could take maximum advantage in the process of de-bugging their new model ready to be launched into the rest of the world. Those two aspects are the line and the trim.

The styling line of the 240Z was everything and more that could be expected from a $3,600 sports car. The interior was almost

The simulated wood grain steering wheel of the original 240Z proved to be slightly slippery and was probably the first thing many a true sporting enthusiast might have changed about the car, though the gear lever knob was real wood.

as good, and was certainly good enough to attract sufficient customers to buy every single car that landed on American shores in the first year almost before it left Japan. The seats were the most noteworthy feature, as they were well shaped and clearly designed to hold the occupants firmly in place. They had adjustable backrests and – not yet common in sports cars of the time – integral head restraints. Head restraints at that time were generally only fitted to rally cars or as extras. Even Alfa Romeos, Aston Martins, Jaguars and Porsches had not yet moved to full-height seats with integral headrests, though some Ferrari models did have adjustable head restraints.

The dashboard of the 240Z was a one-piece moulded unit, into which were installed the instruments, switches and controls, as well as the glove box on the passenger side and the air vents at its extreme ends. The moulding was in a semi-matt grained finish and incorporated five cowled apertures for the instrument dials. The two largest apertures,

Datsun 240Z Two-seater

Construction Integral all-welded steel body and chassis unit

Engine
Crankcase Integral crankcase and cylinder block in cast iron
Cylinder head Cast alloy with steel valve guides and valve seat inserts
Type Six, in line
Compression 9.0:1
Cooling Liquid, with pump circulation and fan assistance
Bore and stroke 83mm × 73.7mm
Capacity 2,376cc
Main bearings Seven, shell-type
Valves Two per cylinder, overhead cam actuation
Fuel supply Diaphragm type mechanical pump and 2 SU-type carburettors
Power output 150bhp at 5,600rpm
Torque Rating 146lb/ft at 4,400rpm

Brakes
Type Servo assisted hydraulic with discs at front and drums at rear
Sizes 10.67in discs, 9.0 × 1.6in drums

Transmission
Clutch type Single dry plate, hydraulic actuation
Gear ratios Manual 11.92, 7.39, 4.77, 3.36:1
 Automatic 8.70, 5.16, 3.54:1
Final drive Hypoid bevel 3.36:1 or 3.54:1

Suspension and Steering
Front MacPherson strut, trailing link arm and tubular shock absorbers
Rear Chapman strut with coil springs and tubular shock absorbers
Steering type Rack and pinion with 2.7 turns lock-to-lock
Wheels Pressed steel 4.5J × 14
Tyres 175SR-14 radial

Dimensions
Overall length 157in (3,988mm)
Overall width 61in (1,550mm)
Overall height 51in (1,285mm)
Wheelbase 91in (2,305mm)
Track (front) 53.3in (1,355mm)
Track (rear) 53.1in (1,350mm)

directly in front of the driver, were for the 260km/h (160mph) speedometer, incorporating a six digit distance recorder and a trip meter, and the tachometer, which read up to 8,000rpm. On the top of the dashboard, above the centre console, were the other, smaller, instrument apertures, the first housing a water temperature and oil pressure gauge, the second an ammeter and fuel gauge and the third a clock.

Warning lights on the dashboard included direction indicators, high beam, handbrake and hazard warning lights, and there were switches for the hazard warning lights, the rear screen heater and two-speed wipers. There was also a cigar lighter. The centre console carried the radio, the heater and ventilation controls and fan switch. Below that was an extension of the console that extended horizontally above the transmission tunnel. Here were positioned the ashtray, the gear lever, a pair of levers controlling the choke and hand throttle and an oddments tray. The door trim was in vinyl in the same colour as the vinyl trimmed seats and the doors were fitted with armrest door pulls for closure.

The heating and ventilation in the 240Z was quite effective, the heater controls providing direction of flow up to the windscreen, to the interior of the car or both. Ventilation was available through the heating vents (with the heater flap closed) and through the air vents positioned on the extreme left and right ends of the dashboard. Outlets were positioned at the rear through the tailgate, just below the rear window. It is as well that the heating and ventilation were as good as they were, for there were no opening quarter lights at the front of the doors and the rear quarter lights were rigidly mounted, not hinged.

Some reviewers thought the interior to be rather dark and oppressive, because if a dark seat and door colour was chosen everything below the dashboard, except for the door trims, was dark in colour, while only that part of the car above the eyeline was light, in particular the headlining, which was also padded. However, there was a logic to having a dark coloured dashboard and lower interior, not least because the dashboard did not reflect into the windscreen and so was not distracting on the road, quite apart from the fact that the dark carpets did not show footmarks so easily and so were easier to keep looking clean.

While the gear lever knob was genuine wood, the rim of the steering wheel was wood-grained plastic. This drew criticism from some quarters, as there was a feeling that a genuine wood rim wheel might have been provided. But to fit one would have cost quite a bit more, and the designers concluded that splintered wood was potentially far more dangerous than shattered plastic if the steering wheel should impact with the driver's chest in an accident. Leather rimmed steering wheels were not yet widely fashionable, so that option was not even considered.

The covering to the gear box housing was seen by many as a bit 'cheap and nasty' and certainly would have benefitted the car if it had been replaced with carpeting, but it was an attempt to produce a 'trendy' interior.

CRITICISMS AND CURES

The rear view mirror was one other area for nit-picking. It was of the dipping type, with a lever to switch from daytime to night use. The problem was that when it was switched, the area in view changed, which meant that the driver had to fiddle with the mirror to bring back the field of view. Another item for criticism was the way in which the side panels of the transmission tunnel were covered. The material chosen was a quilted vinyl, the quilting effect being applied by heat and making the covering look quite cheap. Many buyers of early cars thought the use of carpeting would have been a much better choice for this task, as it would have maintained a style and would have looked better. Some people went so far as to replace the offending trim with carpet material matching that on the car's floor.

Whatever criticisms were offered, though, reviewers had to take into account that the Datsun 240Z was a car built to meet that almost impossible target of price and performance combined. It had to provide a performance to match at least what was already available on the market in its class and it had to be priced at least in the same bracket, preferably even advantageously. In fact, it gave a better performance package than most of its direct adversaries and was ideally priced to ensure that every single one of the first 16,000 exported to the United States was sold long before the year came to its close – many of them at a premium over the sticker price.

Datsun USA took careful note of all the early criticisms and went to work reporting back to Tokyo all they found and heard. In turn, Tokyo went to work on improvements and quickly came up with a package that included a five-speed gearbox option, wider wheels and different tyres, stiffer suspension ratings and detail improvements. In the process, the Datsun 240Z created a special place for itself in the world's sports car markets even a cult, along the lines of the following for the Corvette, the MGB, the Alfa Romeo 1750/2000GT and many others.

The Chevrolet Corvette was America's only true contender for the title 'sports car' and when pitched against the 240Z was faced with a pretty tough piece of competition.

5 The 240Z in the Market

When Albrecht Goertz was brought back to Nissan he was given the task of creating a completely new car from a clean sheet of paper. There was no embargo on the styling of the car, other than the need to be aware of critical comments about the earlier Datsun sports cars. So he was now able to overcome the criticisms that the SP310 and SP311 were too narrow and too boxy, though they were in fact little different from many European cars of similar engine size and class.

GROUND RULES FOR THE NEW Z

The first rule laid down for the development of Nissan's new sports car was that it had to be in the mould of European sports cars, with their kind of performance, style, road speeds and handling characteristics. This was one very good reason for commissioning a European to produce the designs from which this new car would be born! The second rule laid down was that this new car had to be easy to manufacture in volume and the third was that it had to beat the price of the competition.

The most price-competitive European sports cars in the 1960s were those manufactured in Great Britain and their biggest export market at that time was the same as that of Japan – the United States. So the scene was set – the adversaries were to be Japan versus Great Britain, in the form of Nissan versus all comers. The stakes were high, because upon the success of this confrontation lay the whole success potential of Nissan's strategy of creating a world sports car. They had to win this contest in order to further their reputation outside Japan as a serious car maker and innovator and Nissan knew that their new sports model was to be the spearhead of that campaign.

As with its motor cycle industry, Britain was subconsciously abdicating its tenure of the market that Nissan were about to invade. The British Motor Corporation was running into problems that would manifest themselves before Project HS30 had become the beautiful 240Z. The ill-fated British Leyland, with its totally incompatible divisions, was being thrust upon the world by the politics of the day. Furthermore, the British Motor Corporation had killed off the one car that might have held the Nissan challenger at bay – the Austin Healey 3000. A re-style of that car as a Grand Touring coupé, even with the old pushrod C Series BMC engine, might have stood a chance of holding on to the market. But the car that was to take the place of the big Healey stood no chance.

The MGC, little more than an MGB with a Healey-standard 'C' Series engine stuffed into it, was already an outdated body design that, with the serious weight imbalance that resulted from the engine implant, almost guaranteed its failure. The price of the MG did not help it to sell either, so it was not long before the new model was

The 240Z in the Market

The MGC was British Leyland's feeble attempt to recreate the spirit and sales potential of the big Healey. Its sales record told the Leyland board that it had failed and so Datsun cashed in on the opportunity given to them on a plate.

The TR250 had a stripe across the bonnet and the '250' badge to distinguish it from the TR5.

The 240Z in the Market

The Lotus Elan (above), the TVR Tuscan and Triumph GT6 were all contenders for the crown of the new 'Z' car, the first two being close in performance, but not so close on price, whilst the GT6 was a bit cheaper (in the USA) than the Datsun, but fell quite a bit short on performance. Being a lot less expensive in Great Britain, the GT6 fell out of contention.

withdrawn from the market, leaving a gap that would not be filled until the arrival on the scene of the Datsun 240Z.

The Triumph TR5 (the TR250 in America) was a similar story to that of the MGC – it was in fact a TR4 with a 2.5PI engine, though a little more work went into that product, as the TR6 was to be the TR5's successor, carrying with it much of the mechanical specification of the TR5 into a new body styled by Karmann. But the TR6 would have to compete with the 240Z on its own terms, both cars being new to the market together.

Britain's other offerings in the relatively low-priced sector of the market included the smaller-engined and not entirely reliable Lotus Elan, the Reliant Scimitar, with its 2.5- or 3-litre Ford engine, the TVR Tuscan, also using the Ford engine and the Triumph GT6, with the Triumph 2000 engine. Three were glass-fibre bodied cars, which the Americans were still not ready to trust, and the other, while a coupé, was not to be in the same league as the 240Z, not least for the amount of room for its occupants. Also, only the Lotus had an overhead cam engine, in the shape of the Lotus-Ford Twin cam.

The only car of these that would come close to the proposed price for the Datsun 240Z was the Triumph GT6 and the Japanese project team was confident that they could so specify their new model as to enter the American market at just the right level to make it pretty much a straight fight. With the Datsun's better specification, the market resistance was not expected to be too serious, nor sustained. Almost immediately the 240Z was announced to the world, it was acclaimed as having created its own market slot.

The 240Z in the Market

PUTTING THE Z INTO PRODUCTION

The key target for the production of the 240Z from the beginning was a build rate of 2,000 cars per month. This put the stamp of volume-production on the car and meant that it had to be right from Day One. Once the decision was given to go ahead with production, Nissan was in for either the best sales record it had ever enjoyed in North America or a massive flop. To put the scale of the gamble into perspective, the task ahead meant more sales for the 240Z in its first year than Nissan had enjoyed in sports car sales for the whole of the previous five years put together.

Ford was already treading this path with the Capri, a product of combined design and development between Ford of Britain and Ford of Germany announced in January 1969. The Capri could never be considered as a sports car, nor even a sporting coupé in the purist sense of those descriptions, but it was a 'sporty' four-seat coupé, aimed at the motorist who had an aspiration for a sports car, but either could not afford one or needed a car with four seats.

Production of the 240Z began in the second half of 1969, with the objective of putting cars into strategic locations across the United States, to 'feel' the market. The body shell was, of course, totally new, so factory floor space had to be made available for its production. The policy of adapting existing production items helped with component support to the Z line, though of course manufacturing space for the new engines, the L20 for the Fairlady Z and the L24 for the 240Z, had to be accommodated. Conventional assembly of the Z then took place 'down the line'. Despite scepticism in the American and European market, quality control on production lines at Nissan was such that no problems were anticipated at the assembly stage and few were ever encountered.

There are many interesting comparisons to be made between Ford's philosophy for and approach to their Capri and Nissan's attitude to the 240Z. First, the Capri was a product of two factories, Dagenham and Cologne, whereas the 240Z was to be built in the one plant. Despite the fact that Nissan was Japan's largest car maker, it still was not nearly the corporate size of Ford and despite the fact that both the 240Z and the Capri were gambles for their manufacturers, the mighty Ford was far better able to sustain the cash losses of a product failure than Nissan. But there was much more at stake for Nissan anyway, for its whole image as a serious sports car maker would depend heavily upon the success of this new model. Two Ford factories could easily have eased the embarrassment if the Capri did not make it, simply by each factory agreeing to blame the other in its home territory and getting on with the next thing. But Nissan was promoting their new product as a world sports car and much of their general reputation would suffer, as well their name as a sports car maker being totally undermined, if this new model did not make the grade. As it happens, neither maker had to worry, for the 240Z and the Capri were great successes, without either really encroaching on the other's market share.

Within its first year the 240Z fell just short of its production and sales goals, with something over 23,000 being produced by the end of 1970. Exports represented over 18,000 of those cars, with a notable 17,416 going to North American shores (of which 1,201 found their way into Canada). Everywhere the car was seen it received huge public acclaim, and before the year was out 240Zs were changing hands in the United States at substantial price premiums, such was the reputation it was already creating for itself. The Ford Capri had outstripped those modest production figures with sales

The 240Z in the Market

This factory publicity shot of the Fairlady Z shows little difference, apart from badging, between it and the internationally-marketed 240Z.

The 240Z in the Market

The Sports Car Club of America's Production Class "C" Championship was taken by Datsun with the 240Z in 1970.

limited only to Europe in its first production year, but the 240Z had exceeded even Tokyo's wildest hopes, carving out its own market position on the way and, by the end of 1970, winning the Sports Car Club of America's Production Class C Championship. The enthusiasts of the SCCA have always taken their sports car racing pretty seriously and had relied heavily hitherto on British sports cars for their fun.

The second year of production brought even more spectacular results for the 240Z, with some 45,000 built, over 33,500 going to the American market and almost 3,500 entering Canada. Great Britain's single import of 1970 was followed by another 264 in 1971, volumes doubling in 1972 and 1973 as total production figures peaked in 1972, at almost 66,000, fell back in 1973 with the oil crisis and rose again slightly shortly before the car's successor came on the scene. By 1974, there were over 150,000 Datsun 240Zs in the United States – the Z car had found its market niche.

THE Z ON TEST

The Datsun 240Z had already established itself, through the various international motor shows where it had been seen, as a thoroughly stylish vehicle with a fine specification and a price tag that prospective customers simply could not ignore. This car had all the true hallmarks of growing into a classic. It certainly showed the promise of out-selling its competition if it performed nearly as well as its specification suggested. It was now the task of the Datsun test team to make sure it did perform in the way the designers had intended, or report back with sufficient clarity to ensure that modifications made would achieve their goal.

Design and structure testing came first and the new car was subjected to the usual rigours of simulated cold weather trials, when it was put into a huge refrigerator to take its temperature down to sub-Arctic levels; to rough road surface handling trials, being driven over cobblestones for enormous

The 240Z in the Market

The '73 240Z hadn't changed much externally, but it was becoming clear that a bit more engine size would be needed to offset the effects of emission legislation.

distances; to the process of swinging huge weights into the sides of a car to test its impact resistance; and to propelling examples of the car into concrete walls at various speeds to verify crumple rates and to ensure the integrity of the passenger capsule as far as design limits would allow. Then, the car would go through high temperature tests, to ensure nothing would melt that should not and that the car would still start and run satisfactorily for the average motorist.

Track testing was next and the Z was put through a gruelling programme of high- and low-speed handling tests on all kinds of surfaces – smooth tarmac, water, ice, oily water, cobblestones, concrete and any other surface the designers and testers could throw at it to make sure they had produced

the best possible car that Nissan funds and brains could produce. In all, test cars covered over 250,000km (155,000 miles) in the most exhausting test and development programme Nissan had ever embarked upon in the creation of any new car.

After all the bench tests of engine and transmission, the bounce rig tests of suspension and body unit, the achievement of US Federal Motor Vehicle Safety Standards certification and the track testing programme, it was time to put the 240Z to another form of test. This was a test of its performance against targets on ordinary roads. Now, a more conventional programme would take place, in which the new car would be out on the road in Japan and in the United States. These people really took their

The 240Z in the Market

car seriously and were out to do everything they could to make it not just acceptable in its market place, but positively sought-after.

Among the targets set down for the 240Z included a safe top speed of well over 190km/h (120mph), with a braking performance to match, remembering the old safety adage that you can only drive a car as fast as you can stop it. The engine power output target of 150bhp SAE had been exceeded comfortably on the bench and in test cars on the track, so there would be no problem in that area. Nor would there in the matter of torque output, where 140lb/ft was sought and 148lb/ft achieved, so the 0–60mph target of 9 seconds would also be easily reached. (In the late 1960s times of under 10 seconds were quite rare among mass-produced sports cars. While the 240Z turned in 8.2 seconds for 0–60mph, the Triumph TR250 could only come up with 10.6 seconds. The MGA twin cam of a decade earlier was capable of 9.9 seconds on just 1,600cc, while the MGC, on an engine of almost twice the size, struggled to make 10.1 seconds. Just to really push the point home, the Ferrari Dino 246GT, a two-seater of 2.5 litres engine size and something like four times the price, only pipped the 240Z by 0.3 of a second. So the 240Z really was in there among the best.)

ON THE OPEN ROAD

After all the factory test programmes had been completed and the world had seen the new car at various motor shows, it was now time to face the real test of the 240Z – the opinions of dealers and owners. It was not long after its introduction in November 1969 before the first views of the car appeared in the press. *Car Life* was one of the first American magazines to take a look at the new car and they opened their 'first view' (it was not a road test) with the comment: 'The lights are going on all over Europe, in engineering sections and boardrooms. Datsun has a GT car, with all the latest items and a lower-than-most price.' While no-one in Europe was taking the Japanese seriously, the Datsun was truly a potential winner.

This first impression of the 240Z praised the specification of the car, expressing high regard for the 2.4 litre overhead cam engine. At that time, no figures for peak power, engine speed or torque output were released, though they were suggested as being pretty impressive. The writer was taken with the relatively low compression ratio of the oversquare engine, which suggested a particular concern for reliability. He went on to comment about the disc front/drum rear brake layout, observing that, while many an 'expert' might criticize the Datsun for not having disc brakes all round, the disc/drum configuration certainly seemed to work and gave no hint of struggling, even under severe braking from high speeds.

The interior of the car was also praised, with special comment about the seats which, while not possessed of too much longitudinal adjustment, had a rarely found adjustment for rake – not much, but 10° was better than nothing. No comment was offered on the positioning of the three small dials above the centre of the dashboard, but the near Recaro style seats were very highly praised, both for their comfort and the way they held the occupants in the car. The hatchback design was accepted as part of the sporting coupé concept and the large tailgate, albeit very heavy, was given favourable comment for the ease of access it provided.

Sports Car Graphic was the next American journal to see the 240Z and their review put it through its paces, in company with the

The data table opposite from Sports Car Graphic *tells the results of their road test.*

PRICE
Base (estimated) ....$3500 (POE L.A.)
As tested (estimated) $3500
With options ....None

ENGINE
Type ....In-line 6, water-cooled,
  iron block, alloy head
Displacement ....146 cu. in. (2393 cc)
Horsepower ....150 hp @ 6000 rpm
Torque ....148 lbs.-ft. @ 4400 rpm
Bore & stroke ....3.27 in. x 2.90 in.
  (83 mm x 73.7 mm)
Compression ratio ....9.0 to 1
Valve actuation ....Single ohc
Induction system ....Dual S.U.
Exhaust system ....Iron headers, 5 into 2
Electrical system ....12-volt alternator,
  point distributor
Fuel ....Premium
Recommended redline 7000

DRIVE TRAIN
Clutch ....Single dry diaphragm
Transmission   Gear Ratio   Overall Ratio
  1st Synchro ....3.55 ......11.92
  2nd Synchro ....2.20 ...... 7.39
  3rd Synchro ....1.42 ...... 4.77
  4th Synchro ....1.00 ...... 3.36
Differential ....3.36 ratio

BODY
Type ....Unit steel, 3-door, 2-passenger
Seats ....Front buckets
Windows ....2 manual, no vents
Luggage space ....Rear trunk, 9 cu. ft.
Instruments ..160 mph speedo, 8000 rpm tach
  Gauges: ....temp., oil pressure, amp, fuel
  Lights: ....handbrake

CHASSIS
Frame ....Unit steel,
  Front engine, rear drive
Front suspension ......McPherson strut-type,
  with following arms, tube shocks
Rear suspension ....Chapman strut-type,
  tube shocks
Steering ....Rack and pinion,
  2.75 turns,
  overall ratio 17.8 to 1,
  turning circle 31.5 feet
Brakes ....Front discs; rear drums;
  dual independent systems,
  10.7-in. dia. front,
  9.1-in. dia. rear,
  swept area 310 sq. in.
Wheels ....14-in. dia.
Tires ....Bridgestone 175SR14
  pressures F/R: 28/28 (rec.), 32/30 (test)

WEIGHTS AND MEASURES
Weight ......2320 lbs. (curb), 2515 lbs. (test)
Weight distribution F/R 53%/47%
Wheelbase ....90 in.
Track F/R ....53.5 in./53.5 in.
Height ....50 in.
Width ....61 in.
Length ....157 in.
Ground clearance ....6 in.
Oil capacity ....5.3 qt.
Fuel capacity ....15.9 gal.
Coolant capacity ....8.3 qt.

MISCELLANEOUS
Weight/power ratio
  (curb/advertised) ....15.5 lbs. per hp
Advertised hp/cu. in. 1.03
Speed per 1000 rpm (top gear) .....21.8 mph
Warranty ....12 months/12,000 miles

PERFORMANCE

Acceleration ....0-30 (3.1 sec.), 0-60 (8.2 sec.), 0-100 (23.7 sec.)
  0-quarter mile (15.5 sec., 86.5 mph)

Top speed ....135 mph (est.) at 6000 rpm (hp limited)

Braking ....Distance from 60 mph: 151 ft. (0.79 g av.)
  Number of stops to fade: 6
  Stability: Excellent
  Maximum pitch angle: 2.4°

Handling ....Maximum lateral: 0.72 g right, 0.77 g left
  Skidpad understeer: 5.6° right, 4.2 ° left
  Maximum roll angle: 5.1°
  Reaction to throttle, full: Understeer; off: Oversteer

Speedometer	30.0	40.0	50.0	60.0	70.0	80.0	90.0	100.0
Actual	30.5	41.0	51.0	61.0	71.5	81.5	91.5	102.0

Mileage ....Average: 20.8 mpg
  Miles on car: 1440 to 2650

Aerodynamic forces at 100 mph:
  Drag ....270 lbs. (includes tire drag)
  Lift F/R ....—85 lbs./—95 lbs.

TEST EXPLANATIONS

Fade test is successive maximum g stops from 60 mph each minute until wheels cannot be locked. Understeer is front minus rear tire slip angle at maximum lateral on 200-ft. dia. Digitek skidpad. Autoscan chassis dynamometer supplied by Humble Oil.

The 240Z in the Market

Styling of the Opel GT was really the only thing that recommended it as any kind of contender for the 240Z's market.

MGB and the Opel GT. These cars were in the same price bracket as the 240Z, even if not in the same engine size class. In America, engine size was not as important a consideration as it was in many parts of Europe, where it determines the level of road tax payable in some countries and affects the level of insurance premium payable. So a comparison on the basis of price in the United States was entirely reasonable, for fuel consumption would not be a consideration, and since most American enthusiasts would judge performance and price together, they would buy the best deal they could get.

The opening statement of the review in *Sports Car Graphic* reads: 'Maybe it is wrong to tell you at this point in the story, but the Datsun Z Car won...'. It seems that this car had really caught the imagination of the reviewer almost before the meat of the test had begun! The reviewer made the comment that the MG was, of course, the car of nostalgia, because it was MGs that really brought sports cars to America at the end of the World War Two, with many a soldier bringing home an MG Midget from Europe. He went on to comment that, despite the hardships of poor ventilation, inadequate heating and pre-1950s engineering 'you will love' the MG. Given that the comfort, heating and ventilation of the MGB were all justifiably criticized, the car was commended for its handling and braking, the latter being judged by the tester as the best performing of the three cars on test. Also, the gear change on the MGB was declared the smoothest, but not much else.

The Opel GT was thought to be the best styled car of the three. Quite why is difficult to reason, except that it had just a hint of American flavour to its lines. It might be best described as a classic example of the precursor to the 'jelly mould' styling of much more modern cars, but it could not really have been fairly said to be the best looking of this trio. The handling of the German machine was said to be stable, its braking good and standards of interior comfort significantly better than the MGB (not very difficult to achieve). However, the lack of external access to the luggage compartment and very limited access from inside was a feature that was marked down, as was the ultimate performance of the Opel, though the model had a relatively untuned engine. This car would appeal to those who wanted a sporty looking car without it necessarily being a real sports car.

Then came the 240Z. The Opel GT had been described as 'a step towards the ultimate in $3500 GT cars' – that ultimate being the Datsun 240Z, which had clearly captivated its tester for *Sports Car Graphic*. Described as a little bulky-looking (it was a bigger car than either of the other two), its comfort was beyond praise. The silence of the interior when the car was in motion was just not like a sports car! The ride was extremely good, sure-footed and secure, but not quite as good in the opinion of the tester as the MG when pushed. The heating and ventilation were luxury sedan in mode and the all-round vision outstanding. In all, the tester said, it was 'the Ferrari of under-$10,000 automobiles'. Praise indeed.

A Z FOR THE HOME MARKET

Now that the 240Z was firmly on its way in the export market, Nissan could concentrate attention on the Japanese home market. While the export potential of this car was vital, the home market was equally important to the car's establishment as a truly international sports car. What was more, the Fairlady Z would become the symbol of Nissan's success in producing the first ever mass produced sports car in Japan. It would ultimately also become a highly successful competition machine in its home market, in the guise of the Z432R.

By the late 1960s many Japanese were travelling overseas, on business in pursuit of export markets, as buyers seeking out the best bargains in products or raw materials needed in Japan and as government employees taking up posts in diplomatic missions all over the world. These people were beginning to discover what the rest of the world had to offer its citizens and, with the ever-growing Japanese economy, they wanted more for themselves and their own people at home. The consequence was that Japanese industry adapted, for not only could the industrialists at home see a domestic demand that they did not want to miss, they also saw that the more their home market aligned itself to international demands, the easier it would be for them to expand still further their penetration of export markets.

As was the case in many European countries, road licensing of Japanese cars was based on the cubic capacity of the engine, so it was important that the Fairlady Z, given that its lines would guarantee its being a desirable sports car, was an affordable sports car too. There were not many British sports car in Japan at that time, but it was felt that the first Japanese volume-produced sporting machine had to be competitively priced to justify its existence. The Fairlady Z soon justified itself to such a degree that Nissan felt compelled to produce an even more sportier version for the domestic market.

THE Z432 RAISES THE STAKES

In a process of evolution that was aimed at making the 240Z an ever more desirable car internationally, Nissan decided to produce a lightweight version of the Fairlady Z, to be equipped with an engine that was already available, the S20. This was a twin overhead cam 2 litre unit, highly tuned and equipped with three twin-choke Mikuni carburettors

The 240Z in the Market

The Z432 was a twin-cam, three-carburettor 2-litre variant of the 'Z' car for home consumption, though a few did trickle into Australia, it seems. A view of the engine is shown to the left.

The 240Z in the Market

on one side of the cross-flow cylinder head and a six-branch exhaust manifold on the other. The car was developed as Project SP30 and was to become known as the Z432 ('4' representing the number of valves per cylinder, well ahead of the European and American trend for quad valve engines, '3' representing the number of carburettors fitted and '2' the number of camshafts). The rated output for this engine was 160bhp, which was aided by the fitment of fully transistorized ignition, to guarantee a misfire-free spark to each of the engine's six cylinders, even at the highest engine speeds.

In performance terms the Z432, as built, offered a much more interesting engine to look at, and certainly a more intricate piece of engineering, but it was not able to improve on the single cam 2.4 litre unit of the 240Z for road performance. For example, the 0–60mph time for the Z432 was only 8.9 seconds, measured against the lower time of 8.2 seconds for the 240Z. The engine was coupled to a five-speed gearbox, which so many reviewers and owners in the United States had yearned for as part of the specification for the 240Z. Other 'goodies' included a set of magnesium alloy wheels in place of the pressed steel type fitted to the 240Z.

Clearly, the Z432 was to be the precursor of something even more exciting, and this arrived in the form of the Z432-R. This was an out-and-out racing machine in sports coupé clothing. The Z432-R was intended to be raced and was aimed at the privateers in Japan who might want to buy a locally manufactured machine that could really compete with such imports as the Porsche 911, which was even more horrifyingly expensive in Japan than other countries into which it was imported. But the real purpose behind this very limited production Nissan was to race a car in the shape of the 240Z and achieve the success that would bring increased sales. Not only was that goal achieved, but it set the foundation for Nissan's growth into the international motor racing and rallying scene. It is said that only 362 examples of the Z432-R were built, so if you decide to look for one be warned that your search is likely to be as fruitless as it would be for a Toyota 2000GT – and you will probably have to pay a similar price.

6 Development of the 240Z – and the 260Z Arrives

The first sign of the Datsun 240Z in Europe was heralded with its appearance at the 1970 London Motor Show. As elsewhere, it was greeted with enthusiasm, though perhaps mixed with a little scepticism, as Britain was considered to be one of the nations in which the sports car had first evolved and was also a country in which the creation, manufacture and driving of sports cars were all activities to be taken very seriously. After all, this was the home of such great names as AC, Aston Martin, Austin Healey, Bentley, Jaguar, Lotus, MG, Morgan, Riley, Triumph and TVR. How could this Japanese newcomer possibly invade their market and be a driveable, worthwhile sports car?

The answer to the 'how' was 'carefully, studiously and thoroughly'. All the work that Nissan had put into the creation of their new sports car was now to be put to a real test. Not all British and European sports car enthusiasts thought the Americans were very demanding of their sports cars, though the true enthusiasts there were probably even more demanding than the Europeans. But Nissan were prepared for that and fully expected the test of acceptance to be more rigorous and to take longer than in North America.

THE 240Z 'AMONG THE BIG BOYS'

The 1970 Motor Show at Earls Court saw an impressive array of sports cars on display with, starting at the top end, the AC 428, the Aston Martin DBS, the Ferrari 365GTB and Dino 246GT, the Jaguar E-type and the Maserati Ghibli. In the middle range were the Alfa Romeo 1750 Spyder or 2000 GTV, the Porsche 911 in Targa and coupé forms and the TVR Vixen 2500, while at the popular end of the sports car market were displayed such tasty morsels as the Lotus Elan, the MGB, the Morgan Plus 4 and Plus 8, the Triumph GT6 and the Triumph TR6PI. This was an impressive array of sports cars made by people who had been in the business a long time and who were not about to allow a newcomer to the market to steal their business from under their noses.

Nissan, however, had done their homework very carefully and were certainly not to going to leave the British and European markets without having a serious fight for the business they were after. Part of that homework had been to enter the East African Safari Rally to establish the Datsun name as a serious contender in motoring competition. In 1969 they had taken a class win with a 1600SSS Bluebird; in 1970 that class win had been converted to an overall win in the same event. Now, the 240Z was poised to make its competition debut in Britain, soon after the glitter of the motor show had gone, in the International RAC Rally, one of the toughest road events in the British motor sporting calendar, where it finished 7th in the hands of the leading Finnish rally driver Rauno Aaltonen.

Development of the 240Z – and the 260Z Arrives

For the 1971 season, the Datsun faced the Aston Martin DBS (above) from the top end of Britain's sports cars, whilst Triumph's TR6 (right) put up a healthy performance comparison at the lower end of the price scale.

Having already won the Sports Car Club of America's National Championship for production sports cars in Class C, it was clear that this new invader of the European sports car scene would not go away.

In 1971 some 264 examples of the 240Z were sold in Britain, against just under 900 in Australia and 37,000 in North America. The first cars sold in mainland Europe went to Holland, where three cars tested the

Development of the 240Z – and the 260Z Arrives

The Datsun Violet certainly showed what Datsun could do when it came down to hard competition of the kind faced in the RAC Rally.

Rauno Aaltonen brought this 240Z to seventh place in its first RAC, in 1970.

market's reception of the Z, while 82 went to France in that year. The 240Z had clearly made its mark in Britain and was poised for entry into Continental Europe, taking part in many major sporting events along the way as part of the wider launch of Datsun cars.

Development of the 240Z – and the 260Z Arrives

FACING UP TO THE COMPETITION

The primary competitors for the Datsun 240Z in the British market were the products of the specialists, such as the TVR Vixen 2500, the 3-litre Marcos GT and perhaps the Reliant Scimitar GTE. All these were low-volume production cars. The Ford Capri, designed in the mould of a sports car, but with two extra seats and on sale at a somewhat lower price in basic form than the 240Z, would be viewed as a competitor, produced as it was by a volume manufacturer whose main car production was aimed at family vehicles in various sizes and forms, just as was Nissan's, using the sporting type of car as a flagship to raise the profile of the name and generate sales. Triumph had two contenders – the TR6 sports roadster and the cheaper GT6 two-seat coupé.

Standing the 240Z against just those alternatives of similar engine size and accommodation, regardless of price, makes a most interesting list of adversaries. For example, from the top end came the Ferrari Dino 246GT and the Porsche 911T 2.2 coupé, from the middle range the TVR Vixen 2500 and the Alfa Romeo 2000 GTV, while others 'in the ring' included the Triumph TR6, the Reliant Scimitar GTE, the Marcos 3 litre and the Ford Capri 2600 (available only in Germany). This was quite a formidable list, but not enough to put the 240Z out of the running. Only two of these cars were faster than the Datsun – the Ferrari and the Porsche – though the price differential between them and the Japanese

Reliant's Scimitar GTE and Ferrari's Dino 246GT (overleaf) represented two extremes of competitors on the European scene, the Ferrari being the only true contender on performance, though three times the price of the 240Z.

car was enough to buy another 240Z and still have some change.

It is interesting to compare the sales figures for the 240Z in North America and Great Britain. As the population of the United States is five times that of Britain, with a much lower percentage of sporting motoring enthusiasts, one might have expected the British sales figures for the 240Z to be much higher than they were (less than 1 per cent of United States sales). But there were far more sports cars to choose from in Britain and the insurance grouping and engine capacity rating system was so ludicrously weighted against anything with even the vaguest hint of performance potential, especially if that performer was foreign, that the premiums were three to four times those for a family car.

Given that the insurance premium for a 240Z would not have been far short of that for a Porsche, the price differential for buying the Datsun was still enough to tempt the aspiring Porsche or Ferrari Dino owners who knew they would never be able to afford either of those two cars. It takes quite a car to take on all comers and be beaten only by the two that were rated as possibly the finest sports cars in their class. Despite several of the other cars in the list of potential competitors having larger engines, and some being higher in price, the Datsun beat all the others for power output and maximum road speed and handled in a fashion comparable with any.

OUT ON THE ROAD

On the road, the Datsun 240Z was quick to prove its ability to perform and handle. In terms of roadholding, straight line top speed and interior comfort, its main rivals, certainly on price, were those made in Britain. That was a deliberate strategy, because they were the cars closest to the Datsun on price and so they had to be the first to be dislodged from their position of strength in the United States market. On sales features, the interior fittings and trim

Development of the 240Z – and the 260Z Arrives

Inside the MGB (left) and the Triumph TR6 (right), we see price-comparable interiors against which the 240Z had to compete.

had to be better than on the most popular British cars – the MGB, the Triumph GT6 and the TR6. The ploy was to give the new car the performance of the big Austin Healeys at a price to compare with the MGB.

The newcomer to the 240Z was impressed with the completeness of the specification and the immediate comfort of sitting in the driving seat. The gear lever was in the right place for most, and the steering wheel was located comfortably. Most occupants found that the eye line over the steering wheel was comfortable and the presence of a roof did not intrude on the driver's vision, nor did it make the occupants feel a sense of claustrophobia. Noise levels, such as transmitted road noise, engine noise and wind noise, were better than any of the competitors, though early cars gave a slightly softer ride than was thought ideal and the intrusion of noise was higher.

Much of the development of the 240Z for the road came out of early racing experience, with Pete Brock's BRE team being one of the first to take a pair of these cars on to the race track. His racing team experiences led to the improvement of the car ride by stiffening the suspension, which also conferred the benefit of improved cornering. By the 1972 model, flame retardant materials were used inside the car to increase interior safety, another benefit of racing experience. In the process of improving the car, quality in production was addressed, as were odd quirks that seemed to manifest themselves in individual cars rather than be common to the model, like the tendency to understeer more into right turns than into left ones. The brakes on the early cars tended to gather water, and occasional cars would tend to wander, mostly put down to tyres. Transmission noise was a problem on some cars and many seemed to have a propeller shaft imbalance, which resulted in a vibration at high road speeds. But Nissan listened and spent time and money putting the problems right, so that by the middle of 1972 they had improved the 240Z to being as near fault free as any mass produced car could reasonably expect to be, wherever it was made and for whatever purpose.

Development of the 240Z – and the 260Z Arrives

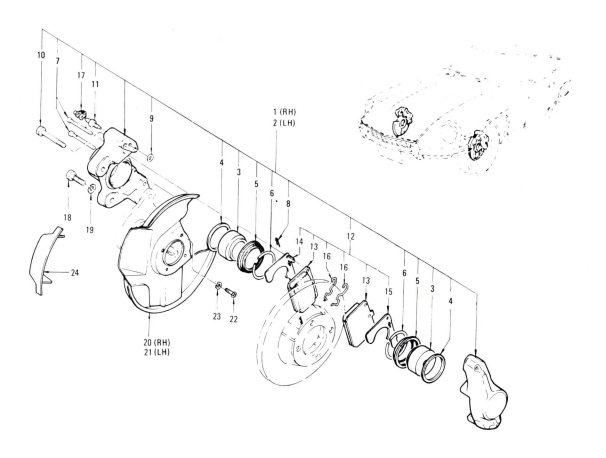

The disc shields on the front brakes of the 240Z generally succeeded in keeping dust out of the brakes, but tended to help keep in water, making handling in the wet a bit tricky on early cars.

DEVELOPING THE 240Z

During the 1971 sales season you could buy your Datsun 240Z with automatic transmission. Before you flinch in horror at the thought of a sports car with automatic transmission, recall that Aston Martin offered their magnificent machinery with an automatic option, as did Ferrari. Many sports cars were becoming available with automatic gearboxes now, so it was quite logical for Nissan to offer the 240Z with that facility. Many sports car enthusiasts have a problem driving a car with manual transmission as the result of a mild disability – they looked upon the advent of an automatic transmission in a sports car as an opportunity to drive an exciting vehicle at last. And many simply wanted automatic transmission for no other reason than comfort.

Road Test in the United States said in its review of October 1971 that the automatic transmission did not harm the performance of the 240Z. This seems a little odd, since such transmissions will always cause some loss of power, simply as a consequence of there not being a direct mechanical connection between the engine and driveline. But as the engine was reported to be identical in specification to that of the manual transmission car, putting out the same 150bhp, the average sporting motorist, as distinct from the tester who wants to wring the last drop of energy from any car, was probably not going to notice a great difference between the two variants of the 240Z.

One area where there would be a noticeable difference in driving technique between the manual and automatic versions of the Z would certainly have been on cornering, where the driver of the automatic version would have to learn the technique of powering a car through bends on the throttle, rather than using slick gear shifts to help the car through. Kicking down or locking into second on an automatic would result in the engine speed being much higher, even close to its limit, compared with a change down to third on a four-speed manual. It would be less noticeable on very tight bends, where a manual would go down to second, making a much closer comparison between the two transmission types in this situation. Also, in a long curve, where a high speed could be sustained, top gear in either mode, manual or automatic, would deliver similar performance. So because the engine specification was identical between the choices of transmission, it might be fair to say that, once the car was running at road speeds, the difference in performance would be minimal, but not identical.

With the self-seeking radio already a standard part of the specification, the only modification to this part of the car would be if the owner was a hi-fi buff and then, perhaps, there would be an argument to change it for something more suitable to that person's taste. But in general terms, this was a sports car and you bought a sports car to derive fun from driving, not listening to music you might find hard to hear at high road speeds anyway. The car's heating and natural ventilation system were very adequate for most conditions in the world, but those owners who lived in extremely hot climates, such as Arizona or Nevada in the United States, or the Middle East, would almost certainly want air conditioning, which was available at extra cost in those markets.

When the 240Z was offered in Britain with a five-speed gearbox it made a huge difference to the car's handling potential. It was a fine car to handle anyway, but American reviewers and owners had said that they had wished it was available with a five-speed gearbox, simply to close the ratios between the gears and allow the driver to get closer to optimum gearing in a much wider range of handling situations. It was likely also to reflect well on fuel consumption, which had been predicted by Nissan to be 25mpg (11.3 litres per 100km). With the five-speed gearbox and moderate road speeds, some road testers claimed it was even possible to push the figure up to nearer 30mpg (9.4 litres per 100km).

During its production life, the 240Z was offered with a varied selection of tyre makes and it seems that the early Bridgestone and Japanese Dunlop specifications did not do the car best justice. As tyre makers in Japan became more familiar with tread compounds suitable for European roads, the Z became more adhesive. Certainly, it could always be said of the 240Z that it clearly telegraphed what it would do if held on a certain line at a certain power setting in a bend or corner, so that it was reasonably easy to rectify a situation and avoid an embarrassing moment in almost

Development of the 240Z – and the 260Z Arrives

The prettiest 240Z without doubt was one fitted with the 'G' nose, as on this example here, belonging to Mr Yutaka Katayama, a man who did much to promote the 'Z' car in North America.

any road or weather conditions. By and large, it was considered by most who encountered it to be a superb car against almost any comers. But Nissan was not ready to rest on its laurels just yet. There were new regulations in the United States to face and there was plenty of natural development left yet in the Z concept.

During its currency in the product range, the 240Z generated quite a range of factory optional tuning components, ranging from full-race engine tuning kit to stick-on 'go-fast' stripes, through-flow exhaust conversions, alloy wheels and extra instruments that did nothing for the performance of the car, but told the driver more about how it was doing what it was doing. Body trim components were also made, including an aerodynamically improved 'G' nose. This nose conversion enhanced the appearance of the car's front end and could well have been the production nose. Short of moving up to the Z432R, with its 250bhp racing mill, there were all sorts of other 'goodies' available to make the 240Z go quicker while retaining its reliability.

THE 260Z ARRIVES

By the end of 1973, exhaust emission limitations had become more stringent in the United States, and since this was their most important market Nissan had simply been compelled to comply. The consequence was that the power output of the original L24 engine had been reduced by stifling the original exhaust output, and they had to do something to restore the power to weight ratio of the original car. The 1974 solution was to enlarge the engine to almost 2.6 litres (2,565cc), lower the compression ratio just a little to 8.3:1 and re-design the camshaft profiles to give a power output of 162bhp at 5,600rpm and provide maximum torque of 152lb/ft at 4,400rpm. So the engine speeds for maximum power and torque were retained, though the lower compression ratio and increased stroke went together to give a cleaner burn and cope with the restrictions of the new compliant exhaust system. Carburation remained two Hitachi made SU type units, though they continued in service to give some of the problems that the carburettors on the

Development of the 240Z – and the 260Z Arrives

late 240Zs gave as a consequence of the emission control fittings – in fact many owners in Britain fitted the early 240Z-type carburettors to eliminate the flat spots and occasional power loss suffered with the new model.

The styling of the 260Z two-seater was basically the same as its predecessor. Such was the success of the line of the 240Z that it was thought unnecessary to re-design the car as a whole. The electrical system was deemed ready for revision, though, so was updated to provide for heavier loads and to limit water ingress and corrosion. The battery was enlarged to a 60 ampere hour unit from 50, while the alternator capacity was similarly increased. Twelve fuses were in the dashboard-mounted fuse box instead of the former ten and while the wheels remained the pressed steel type with the rather unattractive disc trims, certainly in Britain the optional alloy wheels were very popular and improved the appearance of the car significantly. Tyre size was increased to 195/70-14 from 175-14 to take the extra weight of the new model, as it had gained 140lb (64kg) on the 240Z, and the final drive ratio on the new model was 3.70:1.

Driving the 260Z was little different from its predecessor, the top speed being comparable with to the original 240Z at 127mph (204km/h) – against the 125mph (201km/h) quoted by many testers. The 0–60mph time was reduced a little, as was the standing quarter mile. But the new car was even more a 'Grand Touring' car than the 240Z had been. Like the 240Z, it was spacious inside and there was little different in trim or interior fittings. The instruments remained the same, the seats remained fundamentally the same, still with leather-look vinyl trim, though the steering wheel now had a simulated leather trimmed rim instead of the simulated wood of its predecessor, along with the gear shift lever. Other aspects of interior trim were slightly improved, though it was to be a couple of years further on before cloth trimmed seats would arrive.

The 260Z offered for sale in Europe was now somewhat different from its United States counterpart, as the emission control equipment on the United States version set it apart in ultimate performance, fuel consumption, weight and interior trim. If the

Apart from badging, there wasn't an awful lot to tell you this was the 260Z.

Datsun 260Z Two-seater

Construction Integral all-welded steel body and chassis unit

Engine
Crankcase Integral crankcase and cylinder block in cast iron
Cylinder head Cast alloy with steel valve guides and valve seat inserts
Type Six, in line
Compression 8.8:1
Cooling Liquid, with pump circulation and fan assistance
Bore and stroke 83mm × 79mm
Capacity 2,565cc
Main bearings Seven, shell-type
Valves Two per cylinder, overhead cam actuation
Fuel supply Diaphragm type mechanical pump and 2 SU-type carburettors
Power output 162bhp at 5,600rpm
Torque rating 152lb/ft at 4,400rpm

Brakes
Type Servo assisted hydraulic with discs at front and drums at rear
Sizes 10.67in discs, 9.0 × 1.6in drums

Transmission
Clutch type Single dry plate, hydraulic actuation
Gear ratios Manual 10.77, 7.03, 4.85, 3.70, 3.18:1
Final drive Hypoid bevel 3.70:1

Suspension and Steering
Front MacPherson strut, trailing link arm and tubular shock absorbers
Rear Chapman strut with coil springs and tubular shock absorbers
Steering type Rack and pinion with 3.1 turns lock-to-lock
Wheels Pressed steel 5J × 14
Tyres 195/70VR-14 radial

Dimensions
Overall length 162in (4,115mm)
Overall width 64in (1,630mm)
Overall height 51in (1,285mm)
Wheelbase 91in (2,305mm)
Track (front) 53.3in (1,355mm)
Track (rear) 53.1in (1,350mm)

Development of the 240Z – and the 260Z Arrives

Inside, the 260Z had a simulated leather-trimmed steering wheel and gear lever knob, abandoning the rather cheap-looking wood-grain plastic of earlier models.

The rear lights were the main giveaway of the 260Z, with the reverse lights set away from the main tail lights.

United States version had air conditioning, as most did, it would add another 50lb (23kg) or so to the weight and reduce the fuel consumption. But the very label '260Z' was almost certain to boost the slight downturn in sales that had shown itself for 1973. Nissan's sports/GT car was riding high, having taken the production sports car concept into volume-production with total success.

AND NOW THE 2+2

At the same time as announcing the two-seat 260Z, Nissan revealed an expansion of the model line in the 260Z 2+2. As with most such token four-seaters, the room in the back was restricted, but here was a car that could extend its ownership and enjoyment to the driver with a young family. This was very sound strategy on the part of Nissan, for it saw in the creation of the 2+2 the prospect of Z enthusiasts – of whom there were now many tens of thousands, with people owning new and used 240Zs – being able to own their favourite model for ten years or more, without having to contemplate switching to more conventional family transport.

With the creation of the 260Z 2+2, the new car sales volume was now expected to increase. This meant, of course, that more new car sales in the Nissan's sports car range would help the used car market, in keeping prices up, as a wider band of enthusiasts who could not afford a new Z could now buy a used model and so the charisma would be spread. These were good tactics, as any motor trader will tell you that new car sales are always bolstered by good resale values. For a long time, Japanese cars suffered a frightening rate of depreciation. In the early days of Japanese cars being sold in Britain, for example, it was quite common for a one year old car to realize less than half its new value. The international success of the 240Z and the introduction of the 2+2 version of the 260Z were to change all that.

The specification of the 260Z 2+2 included the same engine and gearbox as used in the two-seater – a four-speed gearbox in the United States and five-speed in Europe, with a three-speed automatic as an option. Inside, the car was upholstered in leather-look coarse grain vinyl material

Development of the 240Z – and the 260Z Arrives

*Above:
First of the 2+2
'Z' cars was the 260Z,
a fairly obvious
'stretch' of the
two seater.*

*The interior of
the 260Z 2+2 shows
the leather-look vinyl
upholstery and the
general similarity of
it to the two-seater.*

and foam seating, with vinyl door and side panel trims. The dashboard saw the three central instruments positioned lower than on the 240Z, with controls generally similar to the earlier model. The steering wheel rim was covered with simulated leather, as was the gear lever knob. The centre console, sitting between the front seats, extended to within the reach of the rear seat passengers, but now had no lid, as it was impractical to fit one and retain easy access to all the car's occupants.

To accommodate the additional seats, the wheelbase of the 2+2 was 12in

Datsun 260Z 2+2

Construction   Integral all-welded steel body and chassis unit

Engine
Crankcase   Integral crankcase and cylinder block in cast iron
Cylinder head   Cast alloy with steel valve guides and valve seat inserts
Type   Six, in line
Compression   8.8:1
Cooling   Liquid, with pump circulation and fan assistance
Bore and stroke   83mm × 79mm
Capacity   2,565cc
Main bearings   Seven, shell-type
Valves   Two per cylinder, overhead cam actuation
Fuel supply   Diaphragm type mechanical pump and 2 SU-type carburettors
Power output   162bhp at 5,600rpm
Torque rating   152lb/ft at 4,400rpm

Brakes
Type   Servo assisted hydraulic with discs at front and drums at rear
Sizes   10.67in discs, 9.0 × 1.6in drums

Transmission
Clutch type   Single dry plate, hydraulic actuation, or torque converter
Gear ratios   Manual 12.06, 7.56, 4.77, 3,36:1
                      Automatic 8.70, 5.16, 3.54:1
Final drive   Hypoid bevel 3.36:1 or 3.54:1

Suspension and Steering
Front   MacPherson strut, trailing link arm and tubular shock absorbers
Rear   Chapman strut with coil springs and tubular shock absorbers
Steering type   Rack and pinion with 2.7 turns lock-to-lock
Wheels   Pressed steel 5J × 14
Tyres   195/70VR-14 radial

Dimensions
Overall length   181in (4,595mm)
Overall width   65in (1,648mm)
Overall height   51in (1,290mm)
Wheelbase   103in (2,606mm)
Track (front)   53.3in (1,355mm)
Track (rear)   53.1in (1,350mm)

(300mm) longer than that of the two-seater. The roof line was extended and the car lost the crispness of style of the two-seater, but at least the rear headroom was rather better in the 260Z 2+2 than in many competitive 2+2s. For example, the Porsche 911 coupé gave its rear passengers a stoop if they sat in the car for more than a few minutes, to say nothing of the sparsity of its seating compared with the 260Z. The Ferrari 365GT 2+2 and the Aston Martin V-8 were top of the market cars which retained a sporting character and accommodated rear seat passengers in some degree of comfort, while the Ford Capri 3000GXL and the Ford Mustang gave reasonable rear seat space.

With a quoted top speed of 204km/h (127mph) the 260Z 2+2 was no slouch, though a 10-second 0–60 time was significantly below the level of the original 240Z. However, the 'thump-in-the-back' acceleration was thought less essential, because if the kids were in the back they might not appreciate it! The automatic transmission version of the 2+2 did not give quite such a sparkling performance, but it did manage to give the impression of a sporty motor car and it also enabled a much wider cross-section of the buying public to enjoy the fun of a Z.

Creature comforts in the 2+2 were comparable with those of the 240Z and 260Z, though now the driver and front passenger seats were able to recline a little further than before – provided there were no passengers in the back. If either of the front seat occupants were over 6ft (1.8m) tall the space for leg room in the back was so severely limited that only small children would have been comfortable. However, if the front seats were positioned for an average height adult the rear seat passengers would not be uncomfortable. Placing their feet under the front seats, they had a well padded one-piece backrest and almost as much headroom as those in the front. Ventilation was much improved on the two-seaters, in that the rear quarter light window of the 2+2 was now hinged on its front edge, so could be opened either for the faster through-flow of air or for ventilation for the rear seat passengers.

The 260Z and 260Z 2+2 were announced at the Tokyo Motor Show in 1973 and released into the market before the end of the year, finding their way into the United States significantly earlier than Europe. Both models were on sale in Britain in the spring of 1974. The justification of the 2+2 was quickly shown in the sales figures, for the first year's sales produced over 40,000 two-seaters in the United States, supported by over 9,000 2+2s in the same period. Almost 1,400 two-seaters went to Canada that year, with over half as many again sales for the 2+2. Britain got off to a slow start, with only 129 two-seaters reaching Southampton, but surprisingly 320 of the 2+2 model accompanied them. There were also more 2+2s sold than two-seaters that year in Australia, France, Germany and Holland.

The 260Z two-seater was on sale in the United States for only just over a year and was withdrawn from the European market for a period between 1975 and 1977. The market for two-seaters continued to hold for as long as the car was available in the United States, though the whole Z market began to fall in Canada, slipped in Australia through 1975, but then sold in a ratio of over four-to-one in favour of the 2+2 by 1976. In the British market, the year 1973 was the best for any Z, with over 1,100 cars sold, though almost 450 2+2s entered Britain in 1976. So the decision to introduce the 2+2 was clearly justified and laid the foundation for the next in the line of Datsun Z models.

7 Extending the Line – the 280Z and the 280ZX

Keeping abreast of the competition was an essential part of Nissan's strategy in the development of its sports car range. The United States' Federal Motor Vehicle Safety Standards were an ever-moving process and emission control regulations had reduced the performance potential and power output of the Z to a level where another engine size increase proved necessary to keep any kind of competitive edge. Certain structural changes were essential to keep the market share, too.

DEVELOPMENT OF THE 280Z

Over its life, the Datsun 240Z had faced ever-tightening emission controls in the United States, its prime market, and growing pressure for the introduction and enforcement of crash impact standards in the light of accident statistics that were collected and analysed. The first step taken by Nissan to counter the effect of those regulations was to enlarge the engine to 2,565cc and reinforce the bumpers to withstand statutory impact levels. This resulted in the 260Z being notably heavier than the 240Z, in the price of the two-seater increasing to $5,000 and in the acceleration time from 0–60mph rising to 10 seconds. Now, for 1975, the extended bumper version of the Z had gained a further 130lb (59kg) in weight, just under another 200cc engine capacity, a new fuel delivery system, a $1,300 price hike and a new badge that labelled it as the 280Z.

It has been said that the original Goertz styling had not been tampered with in the development of the car from 240Z to 280Z, but that is not quite correct. The 280Z had very much enlarged front and rear bumpers, which were distinctly ugly and totally out of keeping with the trim line of a sports car, but they had to be there to protect the car under the requirements of the Federal pendulum test (the Safety Standard required that a weighted pendulum had to be able to strike the car's bumpers and not cause impact damage to the bodywork of the car). The result on most cars sold in the United States at that time was bumpers that looked as though they were made for larger cars than those to which they were fitted. The other item that was not part of the original Goertz styling was the tacked-on rear spoiler, which was very attractive to some and grotesque to others. However, it is true that the basic line had not been interfered with, except for the 2+2.

The 280Z two-seater was introduced in 1975 for sale in the United States only, as it was intended to retain the 260Z for sale in other markets. Because the carburation of the earlier models, the 240Z and 260Z, had become increasingly difficult to tune for the retention of smooth running in all conditions with emission controls installed, Nissan fitted fuel injection to the 280Z. While this restored the power output to almost 150bhp, it still had a long way to go to restore the power-to-weight ratio, as the

Extending the Line – the 280Z and the 280ZX

The 280Z brought with it the most obvious changes to the line of the 'Z' two seaters, with its larger energy-absorbing bumpers.

280Z two-seater weighed an incredible 520lb (236kg) more than the original 240Z. It is hardly surprising that the 0–60mph time could not equal that of the original model.

THE POWER PACK AND DRIVELINE

The L28 engine in the 280Z was basically still the same Z cylinder block, but now with a bore of 3.4in (86.1mm) – compared with 3.3in (83mm) of the L24 in the 240Z – and a stroke of 79mm, an increase of 0.2in (5.3mm) over the 240Z. These dimensional increases brought extra weight in the pistons and especially in the crankshaft, which accounted for part of the overall weight increase of the car, but the fuel injection installation did much of the rest. It was decided that fuel injection was the best, if not probably the only, way to cure the flat spots that occurred in the carburated, fuel emission controlled L24 and L26 engines. What was more, specific fuel consumption was likely to be better with the injected engine, too, though gearing in the car would influence consumption on the road.

The fuel injection system selected for the 280Z was the Bosch L-Jetronic system, licence made by Diesel-Keiki in Japan under Bosch patents. It was basically the same system as that adopted for use in the BMW 530i and the Porsche 911 in Europe. Developed from the K-Jetronic mechanical system – also used by Porsche on their superb 911SC Carrera 2.7 – this new version was electronically controlled. Instead of feeding fuel continuously to the engine regardless of circumstances (as many other systems had before it) the L-Jetronic relied on a metering signal from a flap in the air intake, which read mass airflow. This signal combined with other factors, such as throttle position, to feed data into the

Extending the Line – The 280Z and the 280ZX

The engine of the 280Z shows it to be a plumber's nightmare!

system computer, which then regulated the period of time that each injector remained open to feed fuel at each revolution of the engine.

With the L-Jetronic fuel injection system in place the old lean surges, hesitations through flat spots, periodic engine stalls and other ills that plagued the later 240Zs and the 260Z seemed to have been eliminated. Cold starting was a problem of the past, as the injection system recognized a cold engine and simply fed more fuel until the engine temperature was up to operating level. And despite the extra 200cc or so the fuel economy of the L28 engine was set to be somewhat better than that of either the emission controlled L24 or the L26. There was one minor set-back, however, and this was with low engine speeds. Throttle response at low engine speeds, or on a light throttle out on the road, seemed to be slow, as the ignition system varied the spark at about 2,700rpm. At that engine speed, the spark advance seemed to 'jump', but once the engine was above that speed, it ran well and very smoothly.

Probably for reasons of economy in manufacture, the new model initially retained the four-speed gearbox in the United States market, despite the fact that a five-speed had been standard in Europe on both its predecessors. However, the ratios were now different from the 260Z, the intention being to keep a smooth transit through the gears to whatever cruise speed the driver was to maintain. The official explanation for the changes was that they were necessary to preserve durability in the rest of the drivetrain. To further aid the driver to wring the most out of the new model, the final drive ratio was now lowered to 3.55:1 – another measure taken to offset that increase on car weight. However, by 1977 Nissan finally relented and fitted the five-speed gearbox to the 280Z, which, of course, greatly improved it. And on California models, the installation of a catalytic converter was mandatory, as part of the new emission restrictions imposed, inhibiting performance still further. Oddly enough though, the catalytic converter was not required elsewhere in the United States.

Extending the Line – the 280Z and the 280ZX

INSIDE AND OUT

Apart from the large impact-resistant bumpers, the 280Z occasionally had a tacked-on plastic rear spoiler. As with many other sports cars of the time, this was used as a styling feature – it certainly was not going to have any material effect on the rear wheel adhesion of the car at the road speeds it would be allowed to be driven at in the United States. The spoiler, when fitted, was of a different colour from the bodywork and so stood out like the sore thumb it was! There is no escaping the fact that the original factory line was very much more attractive without either the big bumpers or the spoiler.

The 280Z two-seater's rear end was unchanged from the 260Z. The earlier model had introduced a change to the tail light cluster, in which the reversing lights were now separate and located just inboard of the main tail lights. The 280Z retained that feature, as well as the re-located front direction indicators, which were now just inside the radiator grille on each side. Being a United States specification car, a pair of repeater lights for the rear flashing direction indicators were positioned on the rear flanks of the 280Z, as on the 260Z (this was not carried over to European specification models). The badging marked the vehicle as a 280Z, but as always was discreet.

Interestingly, the 260Z had been fitted with wider wheels for the European market. It had 5J-14 rims fitted with 195/70-14 tyres (Bridgestone was the original equipment specification), while the United States specification 260Z remained on 4.5-J14s with 170/75-14 tyres. Now, with the introduction of the 280Z, wheel and tyre sizes were standardized between the models and rims on both models were 5J-14 steel types, with 195/70-14 tyres fitted both to the 280Z in the United States and to the 260Z in Europe. Again, the standard tyre specification was Bridgestone, but many a car had its factory-

This view of both 280Zs, the two-seater (behind) and the 2+2, show that the direction indicator repeater lights on the 280Zs were an instant giveaway of the models.

Extending the Line – The 280Z and the 280ZX

The alloy wheels and wider tyres were a cosmetic improvement to the 280Z, as well as improving unsprung weight and roadholding.

fitted tyres changed before going on the road in the hands of its new owner. Goodyear and Firestone did quite well out of these tyre replacements and many cars were fitted with wider rims – 6J or even 7J alloys were favoured, with correspondingly wider tyres.

The fuel consumption quoted for the 280Z was considerably lower than earlier models at 19.8mpg (14.2 litres per 100km), and Datsun decided to restore some of the car's range by fitting a larger tank. Fuel capacity was increased to 17.2 US gallons (65 litres) from 15.8 US gallons (60 litres), stretching the range of the 280Z to a little under 350 miles (563km) for a long drive. The recommended fuel was 91 octane.

The ride of the new 280Z was certainly substantially better for the comfort seeking driver or passenger than that of the 240Z and, as has already been said, the engine response was generally smoother, with the exception of that slight jolt as the spark advanced from 2,700rpm. But at any engine speed over about 3,000rpm the 280Z ran quite smoothly all the way up to peak revolutions or peak road speed. The peak road speed was now back near that of the original 240Z, at up to 127mph (204km/h), largely dependent upon who tested the car and in what driving conditions.

FOIBLES AND FAILINGS

An old Z malady, rear axle noise, continued to plague the 280Z just as it had done the two previous models. When the driver let the throttle off, there was noticeable differential backlash and a clear 'clunk'. This noise was made the more obvious in the 280Z as the fuel injection system had an automatic cut-off on throttle lift at between 3,200rpm and 2,800rpm to reduce the risk of a fuel build-up that could cause an embarrassing backfire. Not only was this rear axle noise an intrusion on the comfort of the car's

Extending the Line – the 280Z and the 280ZX

Datsun 280Z Two-seater

Construction Integral all-welded steel body and chassis unit

Engine
Crankcase Integral crankcase and cylinder block in cast iron
Cylinder head Cast alloy with steel valve guides and valve seat inserts
Type Six, in line
Compression 8.3:1
Cooling Liquid, with pump circulation and fan assistance
Bore and stroke 86.1mm × 79mm
Capacity 2,753cc
Main bearings Seven, shell-type
Valves Two per cylinder, overhead cam actuation
Fuel supply Electronic fuel injection
Power output 162bhp at 5,600rpm
Torque rating 163lb/ft at 4,400rpm

Brakes
Type Servo assisted hydraulic with discs at front and drums at rear
Sizes 10.67in discs, 9.0 × 1.6in drums

Transmission
Clutch type Single dry plate, hydraulic actuation
Gear ratios Manual 4-speed 11.79, 7.38, 4.55, 3.55:1
  Manual 5-speed 11.79, 7.38, 4.55, 3.55, 3.05:1
Final drive Hypoid bevel 3.55:1

Suspension and Steering
Front MacPherson strut, trailing link arm and tubular shock absorbers
Rear Chapman strut with coil springs and tubular shock absorbers
Steering type Rack and pinion with 3.1 turns lock-to-lock
Wheels Cast alloy 5J × 14
Tyres 195/70VR-14 radial

Dimensions
Overall length 173in (4,399mm)
Overall width 64in (1,630mm)
Overall height 51in (1,285mm)
Track (front) 53.3in (1,355mm)
Track (rear) 53.1in (1,350mm)

occupants, it also created a mechanical risk – the prospect of damage to the differential from the abrupt reaction of the differential to throttle lift. Another mechanical noise came from the viscous-drive fan. When it engaged as the engine temperature rose, it made a noise that was positively intrusive.

With the increase in car weight and the enlargement of tyre size on the 280Z, two new problems manifested themselves. The first was very heavy handling at low speeds. Even the 240Z had not been especially light on low speed steering, but the car's exciting performance made this small foible bearable. But when the 280Z came along, weighing more on the front end and with wider tyres, low speed steering demanded a driver with well developed biceps for it to be described as comfortable. Perhaps it had not occurred to Nissan engineers to fit power steering to ease the problem for the average driver. Power steering would have added yet more weight, and the now not-so-modest price tag could have put the 280Z into a different price bracket and might have injured sales, though such cars as the Porsche 924 were still far enough away in price not to present a problem. It was really a matter of how the established Datsun sports car fan would perceive the figure they would have to part with for the next model in line, when they considered the $3,500 price tag of the original 240Z and the $9,000 of the 1978 280Z.

Inside the car you would expect that with air conditioning added to the specification, demisting the glassware would not present a problem. Not so with the Datsun 280Z! The clear, easily-to-read controls of the new air control system were excellent, as was the quiet four-speed blower fan. As the driver noticed the windscreen misting up, a quick flick of the controls to clear the screen produced an almost instantaneous response and resulted in a beautifully clear forward view. But now you could not see out of the side windows or to the rear, as the rest of the car had misted up! The reason was that the air conditioning went on to re-circulated air when the demister was selected, whereas if the demister had used fresh air, the whole car might have benefitted from crystal clear glass.

The last of the foibles – for no doubt others could come up with a much longer list – was the effect of suspension changes on the ride of the car. With the pretty significant increase in weight over the original 240Z, there had been a need to beef up the suspension at

A factory shot of the 280Z 2+2, again showing how the bumpers intruded on the design.

the front, so the MacPherson struts were enlarged and the spring rates changed. The car's compliance and rough surface ride were improved significantly, so that the occasionally jerky ride over harsh bumps was almost eliminated. However, the tendency for the front end to wander that had manifested itself in the 240Z seemed now to have returned. Braking had improved substantially with the 280Z – thanks to the greater tyre surface contact area (the surface area of the brake pads and shoes remained the same as on the previous model). That braking performance improvement was a plus, but the extra demands placed on the discs and drums on twisty country roads meant that brake fade was excessive.

Notwithstanding the criticisms levelled at the 280Z, it was a much more comfortable car for long range driving than were its predecessors. With the introduction of the five-speed gearbox in 1977, it became much more fun to drive in testing road conditions.

THE BIG ENGINED 2+2

In the same way that the 280Z was virtually a 260Z with a slightly bigger, fuel injected engine, so it was with the 2+2. In fact, it could be said that whereas Nissan had learned something from each step in the development of the two-seater, this extension of the model range brought little with it to the market place than the bigger, fuel injected, engine.

From the front seat backrests forward, the 2+2 was effectively identical to the two-seater, except that the doors were longer to give access to the rear compartment. And 'compartment' is just about what it felt like for two average adults in the back seats. You needed, just as in the 260Z 2+2, a shoe horn to enter the back, though as long as you were not too broad in the beam you could sit in reasonable comfort over short distances.

Headroom was not bad for a person of average height, but anybody 6ft (1.8m) or more in height had a problem. It was a problem that could have been so easily cured by nothing more complicated than raising the roof line by 2in (50mm). This would have given two very distinct gains to the car: it would have allowed a better curvature on the roof line, so retaining some of the original beauty of the two-seater, and it would have allowed more headroom in the back.

Once inside the rear of the car, the occupants had a limited forward view, but the side view was not too bad compared with many European 2+2s. And as long as the person in front of you was not too tall, then you had just enough leg room to place your feet under the front seat and brush your shins up the front seat backrest without dislocating your knees. To fasten your seat belt was something of a work of art, as there was barely enough elbow room to pick up the buckles and bring the lap-type fittings together around your waist. If there were two of you in the rear seats the other could expect a fairly sharp dig or two in the ribs in the process! It was also a little sad to notice that the seat belts emerged through slots in the vinyl upholstery of what, as in the 260Z before it, was effectively a twin sculpted bench seat.

The engine of the 2+2 was exactly the same unit as that used in the 280Z two-seater, the L28, with the same licence-made Bosch L-Jetronic fuel injection, producing the same throttle 'glitch' as the two-seater did at around 2,700rpm. The standard transmission was a three-speed automatic, but the five-speed manual was available as an option for a quite modest $165 – much better value than the extra $95 you had to pay for the California emissions package.

On-the-road performance of the 2+2 was, by and large, pretty good. While you would expect the 0–60mph time of the larger and heavier car to be more sluggish than the

Extending the Line – The 280Z and the 280ZX

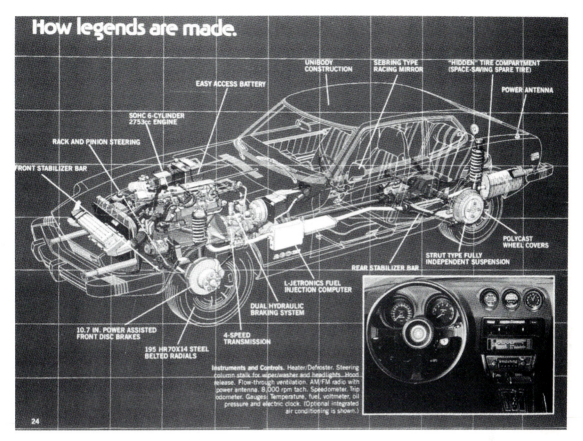

This catalogue drawing of the 280Z 2+2 identifies the line with its immediate predecessor, the 260Z, and also shows where everything is. The mounting pillars for the energy-absorbing bumpers can also be seen, showing how the bumper works.

two-seater, it was still quite good at somewhere between 12 and just under 11 seconds – it depended upon the driving conditions, the driving style of the tester and the individual car under test. One tester managed to wring 20 to 24mpg (11.7 to 14.1 litres per 100km) out of the car on test and a 0–60mph time of 11.7 seconds.

By and large, given that the steering of the two-seat 280Z had been slated for being 'rubbery' and that this characteristic had manifested itself in the 2+2 (since it used the same rack and mountings), the handling of the larger car was said by many to be better than that of the two-seater by virtue of its longer wheelbase. All these criticisms were as valid of the earlier 260Z 2+2 as they were of this model, but it might be fairly said that the novelty of the first 2+2 perhaps allayed the criticisms for a while as people settled down to its presence. And however the critical comments were put, it did have a market, as demonstrated by the level of sales it generated. But, as one person put it in 1978, it was a design 'gone to seed' and was ready for replacement. That replacement was the 280ZX, altogether a horse of a different colour.

Extending the Line – the 280Z and the 280ZX

Datsun 280Z 2+2

Construction   Integral all-welded steel body and chassis unit

Engine
Crankcase   Integral crankcase and cylinder block in cast iron
Cylinder head   Cast alloy with steel valve guides and valve seat inserts
Type   Six, in line
Compression   8.3:1
Cooling   Liquid, with pump circulation and fan assistance
Bore and stroke   86.1mm × 79mm
Capacity   2,753cc
Main bearings   Seven, shell-type
Valves   Two per cylinder, overhead cam actuation
Fuel supply   Electronic fuel injection
Power output   162bhp at 5,600rpm
Torque rating   163lb/ft at 4,400rpm

Brakes
Type   Servo assisted hydraulic with discs at front and drums at rear
Sizes   10.67in discs, 9.0 × 1.6in drums

Transmission
Clutch type   Single dry plate, hydraulic actuation
Gear ratios   Manual 5-speed 11.79, 7.38, 4.55, 3.55:1, 3.05:1
Final drive   Hypoid bevel 3.55:1

Suspension and Steering
Front   MacPherson strut, trailing link arm and tubular shock absorbers
Rear   Chapman strut with coil springs and tubular shock absorbers
Steering type   Rack and pinion with 2.7 turns lock-to-lock
Wheels   Cast alloy 5J × 14
Tyres   195/70VR-14 radial

Dimensions
Overall length   186in (4,714mm)
Overall width   65in (1,650mm)
Overall height   51in (1,305mm)
Wheelbase   103in (2,606mm)
Track (front)   53.3in (1,355mm)
Track (rear)   53.1in (1,350mm)

Extending the Line – The 280Z and the 280ZX

These three pictures show the lines of three of the development models which led to the final styling of the 280ZX.

ENTER THE 280ZX

The march of time is inexorable and progress dictates that even the best in mass-produced sports cars – about 150,000 280Z two-seaters were built – has to have a successor. With oil crises come and gone, leaving the price of crude oil at a level only encountered in consumers' worst nightmares, the cost of running motor vehicles was now a more serious consideration than ever before, even in the oil-rich economy of the United States. The 'gas guzzler' was no longer something that Americans – or people anywhere else in the world – could easily afford. Out of this

The 280ZX 2-seater was quite different from the earlier 'Z' cars, both in line and interior appointment levels (above left), to aim at a higher market level.

Datsun 280ZX Two-seater

Construction   Integral all-welded steel body and chassis unit

Engine
Crankcase   Integral crankcase and cylinder block in cast iron
Cylinder head   Cast alloy with steel valve guides and valve seat inserts
Type   Six, in line
Compression   8.3:1
Cooling   Liquid, with pump circulation and fan assistance
Bore and stroke   86.1mm × 79mm
Capacity   2,753cc
Main bearings   Seven, shell-type
Valves   Two per cylinder, overhead cam actuation
Fuel supply   Electronic fuel injection
Power output   135bhp at 5,200rpm
Torque rating   149lb/ft at 4,400rpm

Brakes
Type   Servo assisted hydraulic with discs at front and drums at rear
Sizes   9.92in diameter front and 10.59in diameter rear discs

Transmission
Clutch type   Single dry plate, hydraulic actuation
Gear ratios   Manual 12.29, 7.68, 4.84, 3.70, 3.19:1
    Automatic 8.7, 5.16, 3.54:1
Final drive   Hypoid bevel 3.70:1 or 3.54:1

Suspension and Steering
Front   MacPherson strut with leading link and tubular shock absorbers
Rear   Semi-trailing arms with coil springs and tubular shock absorbers
Steering type   ZF-type recirculating ball with hydraulic power assistance
Wheels   Cast alloy 5.5J × 14
Tyres   195/70HR-14 radial

Dimensions
Overall length   174in (4,420mm)
Overall width   67in (1,689mm)
Overall height   51in (1,295mm)
Wheelbase   91in (2,319mm)
Track (front)   54.9in (1,394mm)
Track (rear)   54.7in (1,389mm)

situation, Nissan saw a golden opportunity and seized it. They would now steer the Z towards a wider market of users who wanted style combined with relative economy. For Nissan, this meant greater concentration on the 'Grand Touring' aspect and a little less

Extending the Line – The 280Z and the 280ZX

emphasis on the 'Sports Car' for the young blood. Inevitably, the target age group for the next car in the line was perceived to be higher and the car would be aimed at a more stable sector of the population – not an image nursed by the majority of existing owners and enthusiasts.

In charge of the Nissan United States product development team at the time was an Englishman named Peter Harris. His team concluded by 1977 that the Z had moved up-market enough to prove that it was capable of leading its own sales and so, a year later, the 280ZX was to appear. This car deliberately took the type further up-market, to move its image away from the potential competition of Toyota and Mazda, both of whom now had cars in the market that would have been direct competitors in the two-seater market.

The concept was to take the existing and successful engine and driveline of the 280Z and convey it to an entirely new two 2+2 design that would attract the buyer of a more luxurious 'sporty' car, rather than a 'sports' car. This sector of the market was

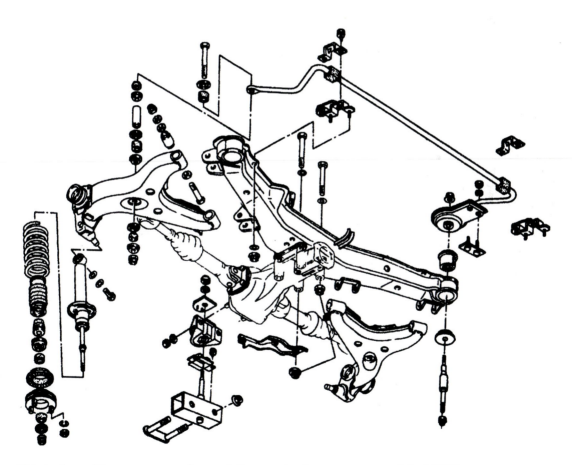

This is the trailing arm suspension and disc rear brake arrangement of the 280ZX.

Extending the Line – the 280Z and the 280ZX

	Datsun 280ZX 2+2
Construction	Integral all-welded steel body and chassis unit
Engine	
Crankcase	Integral crankcase and cylinder block in cast iron
Cylinder head	Cast alloy with steel valve guides and valve seat inserts
Type	Six, in line
Compression	8.3:1
Cooling	Liquid, with pump circulation and fan assistance
Bore and stroke	86.1mm × 79mm
Capacity	2,753cc
Main bearings	Seven, shell-type
Valves	Two per cylinder, overhead cam actuation
Fuel supply	Electronic fuel injection
Power output	135bhp at 5,200rpm
Torque rating	149lb/ft at 4000ppm
Brakes	
Type	Servo assisted hydraulic with discs at front and rear
Sizes	9.92in diameter front and 10.59in diameter rear discs
Transmission	
Clutch type	Single dry plate, hydraulic actuation
Gear ratios	Manual 12.29, 7.68, 4.84, 3.70, 3.19:1
	Automatic 8.7, 5.16, 3.54:1
Final drive	Hypoid bevel 3.70:1 or 3.54:1
Suspension and Steering	
Front	MacPherson strut with leading link and tubular shock absorbers
Rear	As 280ZX
Steering type	ZF-type recirculating ball with hydraulic power assistance
Wheels	Cast alloy 5.5J × 14
Tyres	195/70HR-14 radial
Dimensions	
Overall length	179in (4,539mm)
Overall width	67in (1,689mm)
Overall height	51in (1,300mm)
Wheelbase	99in (2,520mm)
Track (front)	54.9in (1,394mm)
Track (rear)	54.7in (1,389mm)

the 'personal car' buyer, who wanted electric windows, power door mirrors, air conditioning – and even cruise control, anathema to the true sports car enthusiast. This car was targetted more at the Thunderbird/Cadillac owner than the former Austin Healey owner, as a consequence of which it lost some of the 'snap' of the original 240Z

A truly magnificent example of a 240Z in non-standard colour and with non-standard wheels, though the wheel centres do have the proper little 'Z's in them. The colour suits the car extremely well, accentuating the flowing lines of Japan's most successful sports car.

A variation on the original theme was, of course, the Japanese home-market 'Fairlady Z'. This example is of particular interest as it has the 'G' nose and faired-in headlamps, together with extension wheel spats.

The 260Z brought the 2565cc engine to combat the performance stifling emission controls that plagued United States market cars, but in Britain, the larger engine without emission control kit was a bonus. Apart from the badging the two models looked alike at a glance.

To accompany the 260Z two-seater, Nissan decided they could expand their market share by offering an additional pair of seats, so the 260Z 2+2 joined the party and was a huge success. In Britain, alloy wheels came as part of the package.

At home in Japan, the 2+2 was known as the Fairlady 'Z' 2/2, but otherwise looked the same as the export 260, though it had the austerity wheels as standard equipment.

It was predictable that, after the success of the 280Z in the United States, combined with the growing popularity of fuel-injected engines all over the world, an updated 280 would become universally available in the form of the 280ZX. Apart from engine specifications and detail differences, this was the sporting two-four that launched Nissan into the eighties. After 1984, the year of this car, the name 'Nissan' appeared on the cars as well as Datsun. The next model in the line would drop the name 'Datsun'.

Would you call a quantity of 'Z' cars a 'clutch'? Perhaps not, but here's a fine collection of 'Z's, with front left a 'G' nosed Fairlady 'Z', and front right a 280ZX. The red car in the centre in Lynn Godber's 240Z, whilst next to it is the Fairlady 'Z' 2/2 and beyond it, the red 260Z 2+2, whilst the car at the rear right is the 260Z two-seater featured earlier.

The 1982 280ZX Turbo Targa was offered on the domestic market as the Fairlady 280ZX. It was an immensely reliable car and was capable of 'mixing it' with many cars costing two or three times its price.

As the L28 six-cylinder engine drew towards the end of its useful life, Nissan was beavering away to produce its successor. These are just two of the many profile sketches which came before the definitive car.

Here is that definitive car, the 300ZX; this one is a home-market Fairlady.

The most exotic 'Z' car yet, the 300ZX Series Z32, shown here in cutaway form. Whilst still being a volume-produced sports car, the 300ZX was now a legitimate match for all but the most expensive exotica.

Inside the 300ZX.

A little bit of nonsense was this Rinspeed 300ZX with prancing horses all over it. In real life the 'prancing horse' would be unlikely to be 'all over' a Z32 series 300ZX.

After winning Road America 1991 and building up an enviable reputation in competition ...

... they finally did it! A Nissan won the Daytona 24 Hours Race. It is seen here at a night pit stop before going on to victory.

design. It was a factor not lost on Albrecht Goertz who, in a press interview, commented that the new car had lost the clean simplicity of line of the original 240Z. He was absolutely right, though the sales success of the new model showed that the 280ZX worked for its target market.

This was not, as many have perceived it, a facelift of the old 280Z. The 280ZX was a completely new car into which had been transferred an existing and reliable engine and driveline. The front suspension retained the MacPherson struts, but with the tension rod running forwards, not to the rear as on the earlier models. The rear suspension came, in essence, from the 810 saloon, which used a trailing arm connected to a combined spring and shock absorber strut that was not of MacPherson origin. At last, disc brakes appeared on the rear hubs, a pretty substantial 11in (269mm) in diameter, while the front discs were now vented 10in (252mm). A proportioning valve controlled the ratio of effort of the rears to fronts and a dual circuit hydraulic system, with tandem master cylinders, provided actuation from the pedal, with servo assistance underfoot. Another 'at last' for many would-be owners was the provision of power assisted steering to offset the consequences of the growth in weight and wider tyres that had come with the development of the Z.

As you might expect, the performance of the new car fell short of its predecessors, running to somewhere between 117mph and 121mph in top speed, with a 0–60mph time reduced to 11.3 seconds at best for the 2+2 (though one ambitious report quoted 9.6 seconds for the two-seater). Fuel consumption improved a little to come in at an average of 22mpg (12.8 litres per 100km). The L28 engine was now putting out some 135bhp at 5,200rpm, rather less than the earlier 280Z two-seater, but it was tuned now more for a smooth ride than 'thump-in-the-back' performance. There were road

The cast alloy road wheels of the 280ZX were much more attractive than the pressed steel examples fitted as standard equipment to most of the earlier models of 'Z' cars.

wheel options of 5J-14 pressed steel or cast alloys of 6J-14 size, the latter being much more attractive and more practical for the 195/70HR-14 tyres. An increase in tyre width came in 1982, with the introduction of 205/70VR section treads on the 6in rim.

STYLING AND DEVELOPMENT OF THE 280ZX

At the time the 280ZX was announced to the world, many reviewers said they thought the 2+2 had a more balanced line than the two-seater. Given that it appeared to be a styling re-vamp of the earlier 280Z 2+2, with a softer line, that view really has difficulty standing up. The two-seater had much more of the appearance of a new design, whether you liked it or not as a Z enthusiast, and seemed to have a much crisper appearance than its larger sibling. Of course, the 2+2 was what Nissan wanted now to push, as the concept was likely to have a wider market appeal than the pure two-seater. Like any other motor manufacturer, Nissan was in the business of making money from making motor cars.

Extending the Line – the 280Z and the 280ZX

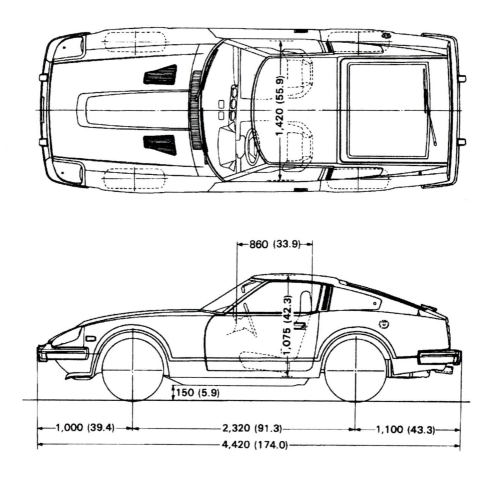

The profile and plan view of the 280ZX show it to be quite a different car from the 240Z of almost a decade earlier.

Inside both variants of the 280ZX, which were designed together from the drawing board up, there was less of a sporting feel to the seats that, thankfully, were now cloth upholstered. They were well designed and held the occupants in place in comfort, but they looked more chunky than you would expect in a sports car. The steering wheel came straight out of a saloon (it could even have come from a Pontiac!) and the instrument panel, while well equipped with dials, was more luxury car than sports car, yet this new car sold more than its predecessors, model for model, and at a higher pricing level.

At the 1979 Frankfurt Motor Show, a new variant of the 280ZX was announced, to go on sale in the United States for 1981. This was the 280ZX Turbo, fitted with a Garrett AiResearch turbocharger to raise the power output of the L28 engine to 180bhp. The show car had another variation for the ZX, in that it had a Targa top. This consisted of two removable glass panels in the roof that created an open car with a roll-bar when

Datsun 280ZX Turbo Two-seater

Construction	Integral all-welded steel body and chassis unit
Engine	
Crankcase	Integral crankcase and cylinder block in cast iron
Cylinder head	Cast alloy with steel valve guides and valve seat inserts
Type	Six, in line
Compression	7.4:1
Cooling	Liquid, with pump circulation and fan assistance
Bore and stroke	86.1mm × 79mm
Capacity	2,753cc
Main bearings	Seven, shell-type
Valves	Two per cylinder, overhead cam actuation
Fuel supply	Electronic fuel injection with turbocharger boost
Power output	180bhp at 5,600rpm
Torque rating	202lb/ft at 2800ppm
Brakes	
Type	Servo assisted hydraulic with discs at front and rear
Sizes	9.92in diameter front and 10.59in diameter rear discs
Transmission	
Clutch type	Single dry plate, hydraulic actuation
Gear ratios	Manual 12.29, 7.68, 4.84, 3.70, 3.19:1
	Automatic 8.7, 5.16, 3.54:1
Final drive	Hypoid bevel 3.70:1 or 3.54:1
Suspension and Steering	
Front	MacPherson strut with leading link and tubular shock absorbers
Rear	Semi-trailing arms, with coil spring and tubular shock absorbers and anti-roll bar (as 280ZX)
Steering type	ZF-type recirculating ball with hydraulic power assistance
Wheels	Cast alloy 5.5J × 14
Tyres	195/70HR-14 radial
Dimensions	
Overall length	174in (4,420mm)
Overall width	67in (1,689mm)
Overall height	51in (1,295mm)
Wheelbase	91in (2,319mm)
Track (front and rear)	54in (1,376mm)

they were removed and the side windows were wound down. The Targa version of the car went on sale first, in 1980, followed by the Turbo engine option a year later.

Whether the Targa was a prettier car than the coupé versions of the current models is a matter for conjecture but, as had been the case with the 2+2, another niche in the

Extending the Line – the 280Z and the 280ZX

The 280ZX Targa was, with the 'T' bar top, an ideal compromise between the now socially unacceptable full drophead and the closed coupé.

market was opened up by its announcement. In performance and interior appointment terms, the Targa was little different from its siblings, being offered in both two-seat and 2+2 forms, but it offered open air motoring for the first time in a Datsun sports car since the days of the SP311 over a decade before. What was more, it was something of a bold move, in view of the general American decline in the popularity of the open car. But it was a bold move which paid off.

The Turbo made a substantial difference to the engine specification, the power unit being labelled L28ET for this application. It had a lower compression ratio, a higher capacity oil pump, larger cylinder head bolts, a modified exhaust system and an electronic engine management system. A larger inlet manifold plenum chamber had a relief valve set at 7.5psi release pressure, there were higher capacity injectors fitted and a larger ignition coil. All to deliver 180bhp at 5,600rpm, which translated to the remarkable 0–60mph time of 6.8 seconds according to one test report. But from the original 240Z, the price tag had grown five-fold!

The Z Turbo in the United States cost a large slice more than the Pontiac Firebird Turbo SS, a little more than the Chevrolet Corvette, but a whole chunk less than a Porsche 924 Turbo – and it was selling well. The car had grown into a wider, heavier machine, but endowed with the turbocharger, it had regained much of the old Z's performance, while most of its potential competitors had become a little more sluggish with each addition of weight brought about by the legislation on emission controls and impact limitations.

8 New Age, New Heights – The 300ZX

The year 1983 was significant in the diary of the Nissan Motor Company Limited, for it had been 50 years since Nissan was formed out of DAT Jidosha Seizo and Tobata. It was also the year in which the internationally known product name was to change from Datsun to Nissan. The name 'Nissan' was already well known in Japan as the product identity, but 'Datsun' had been retained for the export product until now.

The Nissan name was becoming better known in the world outside Japan as a consequence of the company's racing exploits. So the decision was taken to standardize the corporate name as the worldwide product name in this jubilee year and every car would now be called a Nissan. Initially, dual badging was applied, to make sure that everyone knew that a Datsun was a Nissan product. Every car that left the factory had a Jubilee badge affixed to it, which made it even more plain that Datsun and Nissan were one and that the company had been around for quite a long time. This was also the year for the introduction of a new sporting model, now to be known as a Nissan in every market place in the world.

FIRST OF A NEW LINE

The Datsun 280ZX had already been facelifted in 1982 and the 280ZX Turbo had been around for a couple of years, establishing for itself a little of the glory of the 240Z, in that top speed and acceleration performance were better than those figures of the original car – even using automatic transmission! But now it was time to move on. The L28 engine had run its course and it was realized that a new power unit was needed to take Nissan through the 1980s and into the 1990s. The company was now to progress to building the fastest production car ever manufactured in Japan and one of the fastest volume-produced cars in the world.

The new engine in the 300ZX was a V-6, to be known as the VG-30, and it would be installed into a re-vamped 280ZX body unit to make the next step in the Z line. This would take the name of Nissan one step nearer to the Supercar league of sports cars. Offered in naturally aspirated and turbocharged forms, the 300ZX and 300ZX Turbo were available as a two-seater and a 2+2, designed to be almost indistinguishable from each other, the longer rear quarter light being the only obvious difference. Both body versions of the 300ZX were offered as coupé and Targa top types, though there were probably more Targas sold than coupés.

The chassis and body unit of the 300ZX was the same as that used for the 280Z, though it had to undergo some engineering re-design simply to accommodate the new shorter, wider, power unit. Even so, the line of the new model was deliberately styled to bear a strong resemblance to the 280ZX. The

New Age, New Heights – The 300ZX

The Z-31 Series 300ZX retained the lines of the 280ZX, but under the skin was a very different car, powered by the VG-30 2.9-litre V-6 engine, originally in naturally aspirated form, then offered with a turbocharger.

The door handle surrounds, along with the badges and radio aerial mounting, were the only bits of chrome on the 300ZX.

The cast alloy road wheels of the 300ZX suited the style and purpose of the car very well.

Nissan 300ZX Two-seater

Construction	Integral all-welded steel body and chassis unit
Engine	
Crankcase	Integral cast alloy crankcase and cylinder block
Cylinder head	Cast alloy with steel valve guides and valve seat inserts
Type	Six, in line
Compression	9.5:1
Cooling	Liquid, with pump circulation and fan assistance
Bore and stroke	87mm × 83mm
Capacity	2,960cc
Main bearings	Four, shell-type
Valves	Two per cylinder, overhead cam per bank
Fuel supply	Electronic fuel injection
Power output	170bhp at 5,600rpm
Torque rating	175lb/ft at 4000ppm
Brakes	
Type	Servo assisted hydraulic with discs at front and rear
Sizes	9.92in diameter front and 10.59in diameter rear discs
Transmission	
Clutch type	Single dry plate, hydraulic actuation
Final drive	Hypoid bevel 3.70:1 or 3.54:1
Suspension and Steering	
Front	MacPherson strut lower lateral arms and tubular shock absorbers
Rear	Semi-trailing arms with coil springs and tubular shock absorbers
Steering type	Rack and pinion with hydraulic power assistance
Wheels	Cast alloy 6.5JJ × 15
Tyres	P215/60R-15 radial
Dimensions	
Overall length	171in (4,336mm)
Overall width	68in (1,725mm)
Overall height	51in (1,295mm)
Wheelbase	91in (2,319mm)
Track (front)	56in (1,415mm)
Track (rear)	57in (1,435mm)

long flat nose of the 280Z was retained, as was the general character of the roof line. The power bulge on the bonnet of the 280ZX was flattened out, though a small air scoop was positioned on the left hand side of the bonnet on the Turbo. The whole front end of the new model was radically different from the 280ZX, as the round headlamps were replaced with rectangular semi-recessed lamps. The front bumper was integrated into

New Age, New Heights – The 300ZX

the body line, though it was still black, and a black rubbing strip ran along the body sides in line with the bumpers. Another black trim ran below the door line and the trim to the bottom of the front air dam was also black.

The only items of chrome trim on this new car were the door handle surrounds, the badges and the trim to the rear mounted radio aerial. All the window surrounds were black and the windscreen wipers were designed to disappear behind the rear lip of the bonnet when in the parked position. The car was endowed with a large area of glass, the rear quarter light being as large as it could comfortably be within the body line. Gone was the vertical trim on the door pillar of the 280Z, to be replaced with plain paint to give a cleaner line overall. Inside, the 300ZX had deep, dual colour velour-trimmed seats and a soft vinyl fascia and steering wheel rim designed to give the leather look. The centre console was now raised to serve as an armrest and the car displayed an unashamed luxury that would have been totally out of place on the original 240Z in 1969.

The road wheels were a very stylish and attractive cast alloy design that were of metric size (390 × 150mm) on the naturally aspirated version of the car and international imperial (7JJ-16) on the Turbo. Tyres on the naturally aspirated model were 210/60VR390TDs, while the Turbo wore more recognizably sized 205/55VR16s on the front and 225/55VR16s on the rear. With the much larger tyre sizes and the resultant bigger 'footprint', power assistance to the rack and pinion steering was an essential part of the standard specification. Front and rear suspension were an updated version of the 280ZX, though they were tweaked slightly and had the addition of remote control shock absorber adjustment from inside the car, with three alternative damper settings.

POWER PACK AND DRIVELINE

The new engine was a 2,960cc V-6 and fuel injection was used to feed both the naturally aspirated and turbocharged versions. A relatively high compression ratio of 9.5:1 was used on the naturally aspirated engine, which produced 170bhp (now rated to DIN standard, not SAE) at 5,600rpm. This unit had been developed with saloon cars in mind as well as for the new sporting Nissan, for the production numbers of the car would certainly not have justified a whole new engine just for that one type. The turbocharged engine was developed specifically for the sporting model, though, and this engine, with a lower compression ratio of 7.4:1 turned in a very respectable 228bhp. Self adjusting hydraulic tappets helped to keep the engine reliable, maintaining the constancy of its power output.

The standard transmission pack for the 300ZX was a five-speed gearbox, with ratios wide enough to cope with modern traffic conditions, but close enough to give the kind of performance a buyer would expect. With a kerb weight in excess of 3,000lb (1,361kg) this was the heaviest Z ever built – and the

The heart of the matter – the VG-30 engine.

Nissan 300ZX 2+2

Construction	Integral all-welded steel body and chassis unit
Engine	
Crankcase	Integral cast alloy crankcase and cylinder block
Cylinder head	Cast alloy with steel valve guides and valve seat inserts
Type	Six, in V formation
Compression	9.5:1
Cooling	Liquid, with pump circulation and fan assistance
Bore and stroke	87mm × 83mm
Capacity	2,960cc
Main bearings	Four, shell-type
Valves	Two per cylinder, overhead cam per bank
Fuel supply	Electronic fuel injection
Power output	170bhp at 5,600rpm
Torque rating	175lb/ft at 4000ppm
Brakes	
Type	Servo assisted hydraulic with discs at front and rear
Sizes	9.92in diameter front and 10.59in diameter rear discs
Transmission	
Clutch type	Single dry plate, hydraulic actuation
Final drive	Hypoid bevel 3.70:1 or 3.54:1
Suspension and Steering	
Front	MacPherson strut with lower lateral arms and tubular shock absorbers
Rear	Semi-trailing arms with coil springs and tubular shock absorbers
Steering type	Rack and pinion with hydraulic power assistance
Wheels	Pressed steel 6.5JJ × 15
Tyres	P215/60R-15 radial
Dimensions	
Overall length	175in (4,434mm)
Overall width	68in (1,725mm)
Overall height	52in (1,310mm)
Wheelbase	99in (2,520mm)
Track (front)	56in (1,415mm)
Track (rear)	57in (1,435mm)

most expensive. But with the five-speed gearbox it was possible to propel the 300ZX Turbo to 60mph in under 7 seconds and then on, it was claimed, to a 155mph (249kph) top speed. This suggested that Nissan had just about reached the Supercar class.

The automatic transmission fitted to the 300ZX was of Borg Warner type, but now

Nissan 300ZX Turbo Two-seater

Construction: Integral all-welded steel body and chassis unit

Engine
Crankcase	Integral cast alloy crankcase and cylinder block
Cylinder head	Cast alloy with steel valve guides and valve seat inserts
Type	Six, in V formation
Compression	7.4:1
Cooling	Liquid, with pump circulation and fan assistance
Bore and stroke	87mm × 83mm
Capacity	2,960cc
Main bearings	Seven, shell-type
Valves	Two per cylinder, single overhead cam per bank
Fuel supply	Electronic fuel injection with turbocharger boost
Power output	200bhp at 5,200rpm
Torque rating	227lb/ft at 3600ppm

Brakes
Type	Servo assisted hydraulic with discs at front and rear
Sizes	9.92in diameter vented front and 10.59in diameter rear discs

Transmission
Clutch type	Single dry plate, hydraulic actuation
Gear ratios	Manual :1
	Automatic :1
Final drive	Hypoid bevel 3.54:1

Suspension and Steering
Front	MacPherson strut with lower lateral arms and tubular shock absorbers
Rear	Semi-trailing arms with coil springs and tubular shock absorbers
Steering type	Rack and pinion with hydraulic power assistance
Wheels	Cast alloy 6.5JJ × 15
Tyres	P215/60R-15 radial

Dimensions
Overall length	171in (4,336mm)
Overall width	68in (1,725mm)
Overall height	51in (1,295mm)
Wheelbase	91in (2,319mm)
Track (front)	56in (1,415mm)
Track (rear)	57in (1,435mm)

endowed with lock-up devices and an electronic control system which provided two automatically selected shift programmes. The theory was that this would take some of the thinking out of handling the car with automatic transmission. (But if you needed

New Age, New Heights – The 300ZX

not to think while driving, then you needed to be in something other than a Nissan 300ZX – preferably with someone else driving!) It was said that the performance of the automatically transmitted 300ZX was much the same as the manually transmitted car, though road test performances suggest that its 0–60mph time and top speed both fell somewhat short of the figures quoted for the five-speed Turbo or naturally aspirated models.

ON THE ROAD

Comfort, safety and driver gimmicks in the 300ZX were of the highest order yet and, by and large, performance in any of the car's forms was certainly adequate for the vast majority of those people who buy one. In the United States the VG30 engine in the Turbo – even with all the constrictions of emission control equipment and a catalytic converter ensuring that exhaust back pressure did its bit to reduce power output – still turned in a respectable 205bhp. This translated on the road to a 0–60mph time of 7.3 seconds and a theoretical top speed of 135mph.

In Britain, the full 228bhp of the engine was restored and on test, the car turned in a 0–60mph time of 7.0 seconds. The top speed of 140mph (225kph) was a bit disappointing, though, but that might have been much to do with the individual car or the test conditions on the day. But there was no denying that this was the fastest projectile Nissan had ever offered for sale and it was quite a machine for all that.

While priced in the same region as the Porsche 944, the handling and road performance of the 300ZX were said not to equal

The 300ZX was not thought to measure up to all of the characteristics of the Porsche 944 (above left) or the Jaguar XJ-S, but it came close enough to justify its price tag against both.

New Age, New Heights – The 300ZX

The 300ZX SE was the special edition produced to mark the 50th Anniversary of Nissan Motor Company in 1984. Now, its features of body-coloured bumpers, door handles and road wheels, together with its flared wheel arches, sill extensions and front air dam were to be incorporated into the final production version of the 300ZX. This is a 1988 model.

that of many European sports and grand touring cars. It was said to lack the balance of the Porsche 944 or the handling refinement of such cars as the Jaguar XJ-S, though the latter was rarely seen by those interested in real sports cars as being worth much of a second look. The 300ZX was said to have the performance of a supercar, but not quite the manners.

By 1986, the first step in that quest for a true Nissan supercar was taken with the Fairlady 300ZR, a version offered only on the home market in Japan. The new engine used the same cylinder block and crankcase of the original VG30, but now had two camshafts per cylinder head and four valves per cylinder. It was clearly a case of market 'test-bedding' – the new version of the VG30 engine offered in this car was kept at home so that any foibles or service problems could be quickly dealt with before the international community was introduced to the engine. Such was the success of the VG30DE engine that it found its way into the Maxima saloon, but it was never offered on the export market in the 300ZX model.

In 1984, Nissan had displayed a special edition of the 300ZX to mark the 50th anniversary of the company. This version had body-coloured bumpers, door handles and road wheels. It also had mildly flared wheel arches, sill extensions and a smooth air dam to the front. The features of the special edition were now incorporated into the standard production model of the 300ZX in the final facelift of the model. That last facelift was to bring this line of the cars to its end in 1989, making way for what many believe to have been the most exciting model since the announcement of the original 240Z in 1969, just twenty years before.

LAST OF THE LINE – THE NEW 300ZX

Not even Albrecht Goertz could possibly have adversely criticized the new sports car announced by Nissan at the Chicago Auto Show in February 1989. This was truly a Supercar in every sense of the word. On test, it had outrun the Porsche 928S4 at the Nürburgring, lapping in 8 minutes and 40 seconds – Porsche's test driver Gunther

New Age, New Heights – The 300ZX

The new 300ZX (the Z-32 Series, produced out of Programme 901) was a truly elegant sports car, as much of a trendsetter for the Nineties as the 240Z was for the Seventies, though in a different price sector of the market. In tests, this car had stood up to the might even of the Porsche 928, shown here.

Steckkonig had to concede supremacy to a Japanese car for the first time ever. It was as much the result of this test run as anything else that decided the Nissan board to make the launch at Chicago, followed by the Australian and European launches in the autumn, for its time had now come. All versions were on sale in the United States before the end of 1989, the naturally aspirated version in Australia by November and the Turbo 2+2, the only model homologated for Europe, on sale in the spring of 1990, reaching Britain by 1 April.

The chassis and body of the new-generation 300ZX (Project UZ) were completely new and this time, were all-Japanese. Perhaps the lessons of Albrecht Goertz had been well learned and certainly, computers were used extensively in the creation process of critical components. Much reference was made in early information releases, and even brochures, to a Cray Supercomputer, which was used to investigate and verify all the mathematics of the chassis structure and suspension designs, the former being one of the strongest cage structures in sports car manufacture and the latter introducing some quite new concepts to achieve optimum comfort and ride control.

It all began as the result of Yutaka Kume, formerly Nissan's Head of Research and Development, being appointed President of the company in 1985. He saw the need for Nissan to take the initiative in design and technology to place Nissan clearly in the lead in Japan's motor industry, as he

New Age, New Heights – The 300ZX

The 911 Carrera 2 was the culmination of Programme 901 at Zuffenhausen, a single long line of highly successful sports cars, still in production at the time of writing.

perceived that the company was losing ground to its leading rival, Toyota. As a consequence, Programme 901 was instigated, with the objective of placing Nissan at the leading edge of automotive technology by 1990. Interestingly, another Programme 901 had, a quarter of a century before, brought to the world one of the most outstanding sports cars ever conceived – so outstanding that it remains in production yet – the Porsche 911. And it was a Porsche, the 944 Turbo, that had been singled out as the prime target market for Nissan's newest and most revolutionary sports car.

Project UZ was tasked with the objective of creating a sports car for the 1990s, in just the same way as Project HS30 had been charged with producing a sports car for the 1970s. But creating that sports car for the 1990s was a much more difficult task, as the market had become more sophisticated and, perhaps more important, the world had become a place heavy with anti-pollution and safety legislation that had, in later years, much more to do with the politics than science. The design team had to pick their way through the most complicated barrage of regulations in the history of the motor car.

CREATING A SPORTS CAR – AGAIN

Performance was the prime objective of this new car. The second generation 300ZX had to be a performer in the same league as the Porsche 944 Turbo. It had to have passenger safety of a level never before achieved in such a car and its engine had to be so clean as to have no problems in achieving type approval homologation anywhere in the world. The United States market would dictate heavily over issues of interior appointment and, to some degree, the overall style of the car. Surprisingly to some, that market would also dictate a very firm ride compared with other parts of the globe. And it would demand a capability to perform, despite the fact that the new car's performance could never be used to its best on public roads.

With these criteria in mind the design team, under the leadership of Shigeyuki Yamaoka, went to work. The chassis of the car was effectively a cage, constructed of a series of reinforced box sections in zinc-nickel coated with 'Durasteel', a formulation created specifically to be as durable and as corrosion resistant as modern technology

New Age, New Heights – The 300ZX

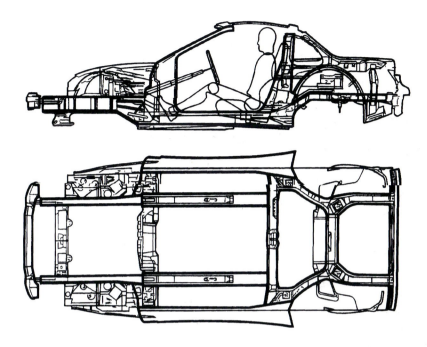

The immensely strong bodyshell/cage of the Z-32 Series 300ZX, produced in 'Durasteel'

would allow. The bulkhead was designed to resist tremendous forces, both transversely and longitudinally, not least because the engine was to be in front, with rear wheel drive. The widespread use of aluminium alloy and plastics was a feature of the body's exterior panelling and trim, to provide the limited yields required under impact.

The styling of the new car used all the tools available to Nissan, from eyeball to computer. While the styling team were anxious not to produce a clone of any other sports car, they were under instruction to create a design that reflected the styling attitudes of the period, as that was clearly what the customer would buy. Many design ideas were sketched by hand, then put through the computer to verify aspects of the concept and examined for suitability. After this, a selection of design ideas that were thought worthy of progressing to model stage, they were modelled and assessed by eye. Further selection resulted in three final designs. These were all modelled in clay to full size and assessed. The lines were again verified by computer and the models refined. Wind tunnel testing cleaned up the lines even further and, after a lot of midnight oil had been burned, the final line of the selected new design was settled upon.

All the time the styling work was progressing, others were beavering away at creating the suspension that would endow the 300ZX with true supercar handling and roadholding, for no-one at Nissan wanted to be told that their new car did not match up to the competition. This new 300ZX bore no resemblance to its immediate predecessor and so its success was very important. The only two aspects of its predecessor that it carried forward were the model name '300ZX' and a development of the VG30 engine. Even the engine, in its quad-cam form from the Fairlady 300ZR, was not left untouched, as it was to have, as an option, twin turbochargers to raise the power output to a higher level than ever before in any Nissan.

New Age, New Heights – The 300ZX

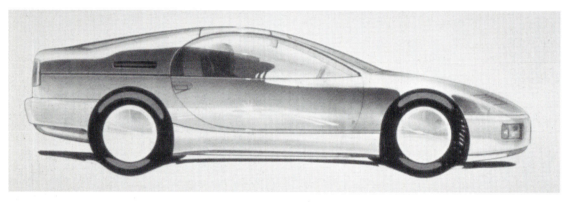

These pictures show styling sketches for the Z-32 model 300ZX before the final line was determined.

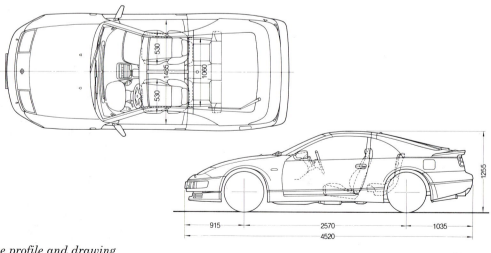

The profile and drawing plan of the 300ZX show its elegant lines to perfection.

Unit: mm

New Age, New Heights – The 300ZX

THE RIDE OF A SUPERCAR

The styling of the newest last Z was up there with the very best – the Porsche 911 Carrera, the Porsche 944 Turbo, the Ferrari 348tb, even the supremely elegant Aston Martin DB7. Now the chassis engineering team had to produce a ride that would take this Nissan into the Supercar category in that same league.

The metal and structural concept of the chassis has already been described, but it is worth examining the design and layout. In planform, the section forward of the B post was a pair of parallel members which supported the floor and seats, as well as the engine at the extreme front end, followed by the gearbox, with a central propeller shaft reaching back to the limited slip differential. At the extreme front end the controlled yield front bumper was attached, while behind the B post were a pair of hoop structures, providing protection for the fuel tank, support for the rear suspension and differential,

Now the 'Z' car was justifiably matched up to the Aston Martin DB7 (above) and the Ferrari 348tb (below).

as well as a mounting for the rear bumper, also of controlled yield design.

The passenger cage was completed by the use of two hoops linked by the T bar roof line longitudinal that supported the inner edges of the glass Targa top panels. The front hoop was created from the front bulkhead and firewall, which extended up to the screen pillars and header rail. The rear hoop formed the rear door shut posts, reaching up and over the seats to provide the rear tailgate header rail. It was a substantially more rigid structure than that of its predecessor and, of course, it needed to be to provide the proper mountings for the car's ingenious multi-link suspension. This massively rigid structure not only provided a sound basis for the use of the lightweight materials of the body panels and sundries, it could also very easily be manufactured in a mass production environment that was to demand a production rate of 5,000 a month.

The multi-link front suspension was designed with the aid of the Cray Supercomputer. The basic double-A arm has added to it a third link, a Nissan innovation that determines the steering axis, freeing the upper arm to maintain the correct geometry for optimum camber angle changes. A bearing in the lower end of the link steers the wheel, resulting in greatly improved stability. Another Nissan innovation is the twisted upper arm which, by virtue of the twist, can increase its effective length for better straight line stability and reduce it for sharper, smoother, cornering.

The rear suspension was also designed as a multi-link system, using two upper links, the spring and shock absorber passing through an eye in one, with a lower A-arm, at the back of which is a lateral link. When cornering or braking loads are applied, the A-arm moves backwards and inwards slightly while the lateral link rotates to the rear. This maximizes stability by adjusting the rear wheel toe-in. Changes in the height of the centre of gravity are eliminated by the upper and lower links.

The design of the rear suspension also provides a facility known as 'HICAS', High Capacity Active Suspension. This provides the advantage of four-wheel steering without the mechanical problems. Sensitive to vehicle speed, steering angle speed and steering angle acceleration, the system alters the

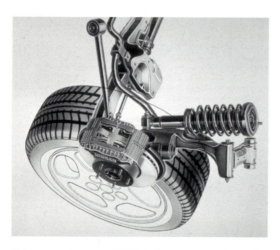

The very clever multi-link front suspension of the Z-32...

...and the multi-link HICAS rear suspension.

Nissan 300ZX (Z-32 Series) 2-Seat Turbo

Construction	All-steel monocoque body/chassis cage with Durasteel panels
Engine	
Crankcase	Integral cast alloy crankcase and cylinder block
Cylinder head	Cast alloy with steel valve guides and valve seat inserts
Type	Six, in V formation
Compression	8.5:1
Cooling	Liquid, with pump circulation and fan assistance
Bore and stroke	87mm × 83mm
Capacity	2,960cc
Main bearings	Four, shell-type
Valves	Four per cylinder, twin overhead cams per bank with VVC
Fuel supply	Electronic fuel injection with turbocharger boost
Power output	300bhp at 6,400rpm (limited to 280bhp on automatic transmission)
Torque rating	283lb/ft at 3,600rpm
Brakes	
Type	Servo assisted hydraulic with ABS
Sizes	4-piston 11in vented discs at front and 2-piston 11.7in discs at rear
Transmission	
Clutch type	Single dry plate, hydraulic actuation
Gear ratios	Manual 11.866, 7.107, 4.807, 3.692, 2.776:1
	Automatic 10.28, 5.70, 3.692, 2.56:1
Final drive	Hypoid bevel 3.692:1
Suspension and Steering	
Front	Independent multi-link with coil springs and tubular shock absorbers
Steering type	Rack and pinion with hydraulic power assistance
Wheels	Cast alloy 7.5JJ × 16 front and 8.5JJ × 16 rear
Tyres	P225/50VR-16 radial front and P245/50VR-16 radial rear
Dimensions	
Overall length	170in (4,310mm)
Overall width	71in (1,800mm)
Overall height	49in (1,250mm)
Wheelbase	97in (2,450mm)
Track (front)	59in (1,495mm)
Track (rear)	61in (1,555mm)

steering angle as vehicle speed increases. This improves steering response and vehicle stability at high speeds and manoeuvrability at low speed, eliminating the feeling of steering numbness often encountered in mechanical four-wheel steering systems.

Nissan 300ZX (Z-32 Series) Two-seater

Construction	All-steel monocoque body/chassis cage with Durasteel panels
Engine	
Crankcase	Integral cast alloy crankcase and cylinder block
Cylinder head	Cast alloy with steel valve guides and valve seat inserts
Type	Six, in V formation
Compression	10.5:1
Cooling	Liquid, with pump circulation and fan assistance
Bore and stroke	87mm × 83mm
Capacity	2,960cc
Main bearings	Four, shell-type
Valves	Four per cylinder, twin overhead cams per bank with VVC
Fuel supply	Electronic fuel injection
Power output	222bhp at 6,400rpm
Torque rating	198lb/ft at 4,800rpm
Brakes	
Type	Servo assisted hydraulic with ABS
Sizes	4-piston 11in vented discs at front and 2-piston 11.7in discs at rear
Transmission	
Clutch type	Single dry plate, hydraulic actuation
Gear ratios	Manual 13.12, 7.86, 5.32, 4.083, 3.07:1
	Automatic 11.37, 6.31, 4.083, 2,833:1
Final drive	Hypoid bevel 4.083:1
Suspension and Steering	
Front	Independent multi-link with coil springs and tubular shock absorbers
Rear	Independent multi-link with coil springs and tubular shock absorbers
Steering type	Rack and pinion with hydraulic power assistance
Wheels	Cast alloy 7.5JJ × 16
Tyres	P225/50VR-16 radial front and rear
Dimensions	
Overall length	170in (4,310mm)
Overall width	71in (1,800mm)
Overall height	49in (1,250mm)
Wheelbase	97in (2,450mm)
Track (front)	59in (1,495mm)
Track (rear)	61in (1,555mm)

POWERING AND DRIVING THE NEW 300ZX

The 60° V-6 VG30 engine used in the original 300ZX was modified so much that it was almost a new engine, though the designation was an extension of the original – VG30DETT in twin turbocharger form and VG30DE in naturally aspirated form. In either case, the engine retained the same displacement of 2,960cc and both the turbocharged and naturally aspirated versions were supplied with fuel by a sequential multi-point electronic fuel injection system with dual plenum intake. Much higher engine speeds were used for maximum power output, both types peaking at 6,400rpm. The naturally aspirated VG30DE produced 222bhp (SAE) on a compression ratio of 10.5:1, while the Twin Turbo VG30DETT gave a massive 300bhp (SAE) on the reduced compression of 8.5:1. This was the most powerful Nissan sports car ever – and all that on premium unleaded fuel, with a catalytic converter in place! The power output on the automatic transmission version of the turbocharged car was reduced to 280bhp, still a pretty fair figure and plenty to propel the car at around 140mph (225kph).

The five-speed manual gearbox had received considerable attention from Nissan's engineers to give it the feel of a true high power sports car. The change lever had a weight built into it below its leather boot and that lever connected to a long and quite substantial selector rod in the gearbox itself. This made for very solid gear changes, and because the lever felt positive and clunked into place there were very few missed gear changes in driving the car. The other factor that helped smooth out the changes was the use of double cone synchronizers, instead of regular single cones, which conferred the additional benefit of a shorter throw on changes.

The VG-30DETT twin-turbo 2.9 litre power pack of the 300ZX.

Four speeds from the automatic transmission, with electronic shift control, gave the driver of the automatic version of the car a much closer feel of the sporting model than previously, so those drivers who did not want manual transmission could still drive a car that gave them the sensation of a true sports car without the manual shifting. The automatic's shift timing and converter lock-up were controlled by electronic signals. Electronics also retarded the ignition timing as each change was made, to reduce input torque and minimize stress. It also had a beneficial effect on fuel consumption, bringing the automatic transmission version of the car closer in fuel consumption to the manual gearbox 300ZX.

Transmitting all this power to the road was approached with every bit as much care as was applied to the development of engine and gearbox. A viscous coupling limited slip differential ensured equality of drive to each rear wheel, so aiding consistent

Nissan 300ZX (Z-32 Series) 2+2

Construction	All-steel monocoque body/chassis cage with Durasteel panels
Engine	
Crankcase	Integral cast alloy crankcase and cylinder block
Cylinder head	Cast alloy with steel valve guides and valve seat inserts
Type	Six, in V formation
Compression	10.5:1
Cooling	Liquid, with pump circulation and fan assistance
Bore and stroke	87mm × 83mm
Capacity	2,960cc
Main bearings	Four, shell-type
Valves	Four per cylinder, twin overhead cams per bank with VVC
Fuel supply	Electronic fuel injection
Power output	222bhp at 6,400rpm
Torque rating	198lb/ft at 4,800rpm
Brakes	
Type	Servo assisted hydraulic with ABS
Sizes	4-piston 11in vented discs at front and 2-piston 11.7in discs at rear
Transmission	
Clutch type	Single dry plate, hydraulic actuation
Gear ratios	Manual 13.12, 7.86, 5.32, 4.083, 3.07:1
	Automatic 11.37, 6.31, 4.083, 2,833:1
Final drive	Hypoid bevel 4.083:1
Suspension and Steering	
Front	Independent multi-link with coil springs and tubular shock absorbers
Rear	Independent multi-link with coil springs and tubular shock absorbers
Steering type	Rack and pinion with hydraulic power assistance
Wheels	Cast alloy 7.5JJ × 16
Tyres	P225/50VR-16 radial front and rear
Dimensions	
Overall length	178in (4,520mm)
Overall width	71in (1,800mm)
Overall height	49in (1,255mm)
Wheelbase	101in (2,570mm)
Track (front)	59in (1,495mm)
Track (rear)	61in (1,555mm)

Nissan 300ZX (Z-32 Series) 2+2 Turbo

Construction All-steel monocoque body/chassis cage with Durasteel panels

Engine
Crankcase Integral cast alloy crankcase and cylinder block
Cylinder head Cast alloy with steel valve guides and valve seat inserts
Type Six, in V formation
Compression 8.5:1
Cooling Liquid, with pump circulation and fan assistance
Bore and stroke 87mm × 83mm
Capacity 2,960cc
Main bearings Four, shell-type
Valves Four per cylinder, twin overhead cams per bank with VVC
Fuel supply Electronic fuel injection with turbocharger boost
Power output 300bhp at 6,400rpm (limited to 280bhp on automatic transmission)
Torque rating 283lb/ft at 3,600rpm

Brakes
Type Servo assisted hydraulic with ABS
Sizes 4-piston 11in vented discs at front and 2-piston 11.7in rear

Transmission
Clutch type Single dry plate, hydraulic actuation
Gear ratios Manual 11.866, 7.107, 4.807, 3.692, 2.776:1
 Automatic 10.28, 5.70, 3.692, 2.56:1
Final drive Hypoid bevel 3.692:1

Suspension and Steering
Front Independent multi-link with coil springs and tubular shock absorbers
Rear Independent multi-link with coil springs and tubular shock absorbers
Steering type Rack and pinion with hydraulic power assistance
Wheels Cast alloy 7.5JJ × 16 front and 8.5JJ × 16 rear
Tyres P225/50VR-16 radial front and P245/50VR-16 radial rear

Dimensions
Overall length 178in (4,520mm)
Overall width 71in (1,800mm)
Overall height 49in (1,255mm)
Wheelbase 101in (2,570mm)
Track (front) 59in (1,495mm)
Track (rear) 61in (1,555mm)

New Age, New Heights – The 300ZX

Nissan 300ZX (Z-32 Series) and Fairlady Convertible

Construction	All-steel monocoque drophead body/chassis cage with Durasteel panels

Engine
Crankcase	Integral cast alloy crankcase and cylinder block
Cylinder head	Cast alloy with steel valve guides and valve seat inserts
Type	Six, in V formation
Compression	10.5:1
Cooling	Liquid, with pump circulation and fan assistance
Bore and stroke	87mm × 83mm
Capacity	2,960cc
Main bearings	Four, shell-type
Valves	Four per cylinder, twin overhead cams per bank with VVC
Fuel supply	Electronic fuel injection
Power output	222bhp at 6,400rpm
Torque rating	198lb/ft at 4800ppm

Brakes
Type	Servo assisted hydraulic with ABS
Sizes	4-piston 11in vented discs at front and 2-piston 11.7in discs at rear

Transmission
Clutch type	Single dry plate, hydraulic actuation
Gear ratios	Manual 13.12, 7.86, 5.32, 4.083, 3.07:1
	Automatic 11.37, 6.31, 4.083, 2,833:1
Final drive	Hypoid bevel 4.083:1

Suspension and Steering
Front	Independent multi-link with coil springs and tubular shock absorbers
Rear	Independent multi-link with coil springs and tubular shock absorbers
Steering type	Rack and pinion with hydraulic power assistance
Wheels	Cast alloy 7.5JJ × 16
Tyres	P225/50VR-16 radial front and rear

Dimensions
Overall length	170in (4,310mm)
Overall width	71in (1,800mm)
Overall height	49in (1,255mm)
Wheelbase	97in (2,450mm)
Track (front)	59in (1,495mm)
Track (rear)	61in (1,555mm)

New Age, New Heights – The 300ZX

stability of the car. Steering was, as with all Z models, by rack and pinion, but now with fully variable power assistance to allow easier handling on the 225/50VR16 tyres, which were mounted on 7.5JJ-16 rims all round on the naturally aspirated car, though the rears were 8.5JJs, supporting 245/45ZR16 tyres, on the Twin Turbo. Brakes were much improved over the earlier 300ZX, with four-piston power assisted 11in vented front discs and two-piston 11.7in vented rears. An electronic anti-lock braking system completed the stopping equipment of this magnificent car, which now weighed from 3,414lb (1,549kg) in two-seat naturally aspirated form to 3,516lb (1,595kg) for the automatic transmission Twin Turbo 2+2.

THE FIRST EVER CONVERTIBLE Z

You might suppose, looking at the 300ZX, that it would never make a successful convertible. But it did and the process was remarkably simple. If you looked closely at the two-seat coupé, then lifted the huge tailgate, you were almost there. That is exactly what Nissan did: they removed the T bar that linked the windscreen header to the B post hoop, taking the glass roof panels with it. They then removed the tailgate and replaced it with a rear deck trunk lid. The B post hoop was retained and padded to prevent the risk of injury. A fold-away canvas hood completed the specification which, in

The car they said Nissan would never build – the 300ZX Convertible: a truly elegant and so simple drophead version of the last 'Z' car.

New Age, New Heights – The 300ZX

A late Z-32 2+2, the car which, in the fullness of time, will surely attract its own following with just as much charisma as the 240Z two decades or so before.

all other aspects, was identical to the two-seater coupé.

First launched as a concept car at the 1991 Tokyo Motor Show, the new model brought such a response that Nissan were persuaded to put it into production the following year. It did not sell by the thousand, but was a worthwhile extension to the range and was an easy conversion to manufacture. First released on to the domestic market in Japan during August 1992, it was later exported to the United States and elsewhere.

THE STORY ENDS

As the 300ZX brought the Z to its silver jubilee, reaching the pinnacle of Japanese sports car concepts, it seems hard to believe that red tape should have eclipsed its sales in Britain in that jubilee year. Withdrawn from the British market at the end of 1994, the official explanation was lack of sales. Impending emission legislation required the car to be fitted with two catalytic converters. There was no room to fit a second catalyser and sales were certainly not sufficient in Britain to justify a major redesign.

While, at the time of writing, the 300ZX continues to be sold in other markets, it seems sad that Japan's finest sports car and the only Japanese car ever to be truly capable of taking on the great names of motoring in Europe's Supercar league should disappear from the market just as it was taking on all that was best. But the British market had the 300ZX for five years and in that time, it made its mark. How history will view it is a matter for time to tell. But it is a fair guess that in a few years time, the Nissan 300ZX of 1989 to 1994 will have as much of a cult market as the 240Z, 260Z and 280Z have today.

9 Facing the Competition

Looking back over a quarter of a century of production, it is interesting to see how the Z grew up with its competitors and moved discreetly up-market in the process. The perennial question 'What is a sports car?' has been asked over and over again. In general terms, the answer to that question must surely be: 'A car capable of higher than average speed for its engine size, with road-holding and braking performance to match, and which is a joy to drive'. Traditionalists have often added the rider that it would also have to be an open car, but that image has been forced to change.

Within the sports car category the 'Gran Turismo' concept has grown in acceptance since the 1950s. Ever since Alfa Romeo fielded its 6c3000CM and Ferrari the 250MM coupés in the 1953 Mille Miglia and Le Mans, there has been a growing mystique around sporting coupés that has helped to make them more popular on the road in the hands of the general public. The Grand Touring car has also been the bridge between the open two-seater and the high performance saloon. With the mood of the late 1960s swinging against open cars in North America, Nissan's prime target market, it was hardly surprising

The Alfa Romeo 6c3000CM and the Ferrari 250MM (overleaf) – two cars which set the trend for the definition of the term 'Gran Turismo'.

Facing the Competition

that the company should go for a coupé to expand its sports car base in that market. This was a notable factor, combined with the build quality and specification of the 240Z, in its rapid acceptance.

It is interesting also to see how Nissan and other manufacturers have developed their cars as legislation has tended to drive specification changes in the whole motor vehicle world. Porsche, for example, has raised the engine capacity of the 911 from 2.2 litres in 1970 to 3.6 litres in 1996 – certainly increasing performance on the way, but amazingly using the same basic crankcase and external engine dimensions. The Ferrari Dino series of models grew from 2.5 litres to 3.5 litres and became the 348, though their twelve cylinder engined range has grown to 5.4 litres. Aston Martin did a similar thing to Ferrari, the principal model being a 5.2-litre V-8, but by re-introducing the DB six cylinder models, in the form of the DB7, re-entered the 3- to 3.5-litre category with their adaptation of the six cylinder 3.2-litre Jaguar engine.

Other sports car makers did the same kind of thing, increasing engine capacity to maintain and occasionally improve the road performance of their vehicles. TVR went from 1,600cc to 3 litres; Morgan went from 1,600cc to 3.5 litres; Ford went from 1,300cc to 3 litres – all in basically the same model as they started with. Datsun went from just under 1,500cc with the SP311 Fairlady to 3 litres in the 300ZX, but in four distinctly different car designs.

STANDING UP TO THE 240Z

Price was the all-important consideration in launching the 240Z, once the styling and engineering design were complete. A competitive price put this car into the market quicker than anything else. Albrecht Goertz's fine lines had to be translated into cost-effective manufacture and this was to be achieved by carefully selecting pressing shapes and metal thickness to hold production costs to a minimum. The old transfer

Facing the Competition

The Datsun 240Z and its Adversaries, 1972–73

Car and Model	Engine Type and Size	Power (bhp)	Gearbox	Max. Speed (kph)	Max. Speed (mph)	0–60 mph (secs)	GB Price
Datsun 240Z	6 cyl in line water cooled OHC 2,376cc	150	4-speed manual	201	125	8.2	£2,389
Alfa Romeo 2000GTV	4 cyl in line water cooled DOHC 1,962cc	124	5-speed manual	177	110	9.6	£4,009
Alfa Romeo Montreal	V-8 water cooled OHC 2,593cc	200	5-speed manual	225	140	7.6	£5,549
Ferrari Dino 246GT	V-6 water cooled DOHC 2,418cc	175	5-speed manual	227	141	7.9	£5,579
Ford Capri 3000E	V-6 water cooled OHV 2,994cc	138	4-speed manual	196	122	8.4	£2,195
Marcos 3 litre	V-6 water cooled OHV 2,978cc	130	4-speed manual	193	120	7.5	£2,793
MGB GT Coupé	4 cyl in line water cooled OHV 1,798cc	95	4-speed manual	166	103	14.6	£1,570
MGB GT V-8 Coupé	V-8 water cooled OHV 3,528cc	137	4-speed manual	201	125	8.6	£2,294
Porsche 911S	Flat-6 air cooled OHC 2,341cc	190	5-speed manual	233	145	8.1	£5,674
Reliant Scimitar GTE	V-6 water cooled OHV 2,994cc	135	4-speed manual	196	122	10.0	£2,398
Triumph GT6 Coupé	6 cyl in line water cooled OHV 1,998cc	104	4-speed manual	172	107	10.0	£1,469
Triumph TR6	6 cyl in line water cooled OHV 2,498cc	142	4-speed manual	195	121	8.2	£1,724
TVR Vixen 2500	V-6 water cooled OHV 2,498cc	108	4-speed manual	179	111	10.6	£2,550

Facing the Competition

machines brought in during the Austin era were replaced by much improved Japanese-made machines, so manufacturing costs of engines were held down. Parts adapted from other production models were employed where practicable, helping further to keep down the final selling price.

Aiming the 240Z at the United States market was clearly a sound strategy for Nissan, as many American servicemen had been, and continued to be, stationed in Japan. The Americans had bought cars and taken them home and some had spread the word, providing a solid market base upon which Nissan could build. As more servicemen returned home, left the armed services and joined their local communities, so the market potential for Japanese sports cars increased. American servicemen were returning from Europe too, bringing home with them examples of European sports cars, but the price factor was now beginning to tell.

The 240Z was quite deliberately pitched at the $3,500 sports car market in the United States, because Nissan knew that their car should out-perform just about all the cars in that bracket on sale at the time – and quite a few priced well above that. Whether there was any pricing subsidy on the part of the manufacturer to aid market penetration is a question that has been asked, but the answer is probably no, because Nissan would almost certainly have been able to control its production costs sufficiently to hold the price where they wanted it, having planned that aspect in the development of the car from concept to production. Also, since the Americans did not have the same insurance grouping problem – biased towards engine size and potential performance – as in Europe, especially Britain, the price of the car was much more relevant. Getting the most for your dollar was what really mattered in this environment.

Prime among the competition for the 240Z was the British contingent, represented by the MGB (the 'C' had gone by now), the Triumph TR250, the Triumph GT6 (a variant of the Spitfire), the TVR Vixen, the Reliant Scimitar GTE and the Marcos 3 litre. In the United States, only the MG and the Triumphs were real contenders, as the TVR and the Reliant were imported in very small numbers, though their statistics are relevant, as the 240Z came to Britain and

The Ford Capri 2600 from Germany brought this car nearer to the performance envelope of the 'Z' car, but it was never seen as a real head-on competitor.

took on that range of cars. Continental European contenders at the top end of the scale included the Ferrari Dino 246 and the Porsche 911T, while the Alfa Romeo GTV 1750 was a middle price contender and the Ford Capri 2600 from Germany at the lower end of the price scale.

The Ferrari Dino 246 was a high priced two-seater, with a V-6 engine of 2.4 litres capacity coupled to a five-speed gearbox. A typical Ferrari, it was an outstanding performer and was certainly no contender in terms of price, but the interesting point about its inclusion in this list is that its acceleration from 0–60mph was only one third of a second faster than the Datsun, which was less than a third of the price in the United States and just over that in Britain. With 175bhp power output and 141mph (227kph) top speed, the Dino was the car you would buy if you wanted a Ferrari, but its performance came at a price.

The Porsche 911T was very nearly an equal match in acceleration time, one tenth of a second faster than the 240Z and about equal theoretically on top speed, though in practice the Porsche could reach 130mph (210kph) before running out of breath, whereas the Datsun was beginning to pant a bit as it topped 125mph (200kph). While the Datsun was said to have a bit of a quiver at high speed, if you pushed the Porsche to its limit and were not ready for what could happen, it could turn about on you and point you in the direction you came from very quickly. Porsches demanded the utmost respect, but if you gave it you had a thoroughly exciting drive, though at twice the price of the 240Z. Porsche had, of course, established an enviable reputation in racing and rallying throughout the world and it was going to take more than a Japanese upstart to break the mould and persuade Porsche afficionados to switch. But again, performance at a price told.

Datsun stood a better chance against Alfa Romeo, though, and the 1750 GTV, while a superb performer in its own class, was also an expensive car to buy. At that time, Alfas had acquired a reputation for corrosion, though the 1750 engine also had a reputation for being unburstable. It was that characteristic, combined with Alfa Romeo's legendary reputation for roadholding and handling, which created 'Alfisti'. But in the United States, where value came in acceleration and top speed, the Datsun was going to stand full square against the Alfa in most areas of performance, with a full second shorter 0–60mph time and at least equal top speed, aided by almost 20bhp more. In addition to performance, the Datsun had the built-in radio and high back seats, but on the down side only four speeds in the gearbox (in the USA). On that basis, the Datsun 240Z was not only in the competition, but was also going to win a few rounds on the way.

The next three contenders – the TVR Vixen, the Reliant Scimitar GTE and the Marcos – can best be described as 'special interest' cars, in that all three were limited production vehicles that tended to sell to enthusiasts for those particular cars. The TVR and Marcos were out-and-out sports cars but neither had the 'legs' of the Datsun, though both were probably capable of out-cornering it on tight British roads. Both were more expensive than the 240Z, though, and while the Marcos was said to be capable of 120mph (193kph) and had a quicker 0–60mph time by three quarters of a second, it was quite a delicate car, with a glass-fibre body on an ash frame. The 240Z was tough and while the Reliant Scimitar, again glass-fibre bodied, also had a good top speed, it lost nearly two seconds on 0–60mph time. The one thing that held Reliant customers was the extra space in the back. It was a 2+2 estate car in concept

Facing the Competition

The Alfa Romeo GTV was an excellent performer, had beautiful styling and was an occasional four seater – but on the downside, cost a lot more than the 'Z' car and quickly seemed to develop galloping body rot.

and that was what sold it, so to buy a 240Z you were going to switch to a sports car.

In the United States the Austin Healey 3000 was a legend, but in the mid 1960s the British Motor Corporation decided it was time to kill the car off. Realizing that it might have made a mistake, BMC decided to squeeze the C Series engine into the MGB, but that was not much of a success either, so the MGC did not last. This gave Nissan their golden opportunity, for the Austin Healey market void was just what they were going to fill. There was an MG on the market, but it was the four-cylinder engined B. Priced in the same $3,500 bracket, the MGB was really no contest from the beginning, with a 0–60mph time of 12.2 seconds, a top speed of only 105mph (169kph) and handling that just could not compete with the 240Z.

The two Triumphs had good followings in the United States, because they had represented excellent value for money. But now, they too had their backs to the wall. The Triumph GT6 was unashamedly a low priced coupé for two. It had a simple and solid pushrod engine propelling it, coupled to a four-speed gearbox that again seemed tremendously durable. Two bucket seats, not a lot of headroom, an attractive line and characteristically 'English' handling that

meant it cornered beautifully for a car of its type, but was a bit of bumpy ride. The GT6 only really found itself in contention with the 240Z in the United States, for the price was only about $500 below the Datsun (though in Britain the GT6 was a lot less expensive). The principal reasons why you might buy a 240Z in the United States were its greater headroom, better seats, radio and, of course, better performance. But in Britain, the 240Z was a positive step up from the GT6.

The TR250 was Triumph's 'real' sports car, successor to the famous line of Triumph TR sports cars that began with the TR2 in the early 1950s. Here was a car that had a reputation for being a real handler. Like all TRs, it was an open two-seater. It had a top speed of 107mph (172kph) and a 0–60mph time of 10.4 seconds. They were fine figures as they stood, but the fact that open cars were declining in popularity at that time and that the TR250 and the 240Z just about matched each other for price meant that the 240Z was destined to eat into a chunk of the Triumph's market.

The Ford Capri had developed into a 2.6 litre V-6 in Germany and a few found their way to America. In both locations it stood below the Datsun on price, but fell short on roadholding, out-and-out performance in

Triumph's GT6 was the nearest pure coupé from Britain to the 'Z' car, but fell short on performance, though not price.

top speed and acceleration and was not a serious Grand Tourer or sports car. As a consequence, the Capri sold to loyal Ford buyers who wanted something better than they had in a saloon car, but it was never a serious contender for a market share of the Datsun 240Z's potential. The 240Z was better appointed, it was a better performer and so became the best selling sports car in North America within a year.

THE 260Z VERSUS THE REST

Many of the adversaries of the 260Z were the same as those facing the 240Z, though Triumph had introduced the TR6 by now and it was a much better car than its predecessor on performance, while the GT6 had gone. The good old MGB was still around and not much changed, though the GT version had now acquired a V-8 3.5-litre engine from Rover, courtesy of British Leyland. The TVR Vixen had grown into the 3000M, the Reliant Scimitar continued as before, while the Ford Capri, still built in Germany, had now acquired extra engine capacity to become the 2800.

Triumph's TR6 had grown into a more acceptable vehicle for the United States, being styled for that market. Some people preferred the lines of the TR4/5 series, but the new car offered a fuel injected engine, a 0–60mph improvement to 8.2 seconds (the same as the 240Z) and a top speed of 121mph (195kph). An optional detachable hardtop made it cosier, but it came up against the 260Z on price and, now, the growing reputation of the Datsun sports car, as well as its much more attractive lines. Another problem that was beginning to beset the British cars was lack of reliability. Warranty claims were building up for the silliest little faults, like body rattles, poor window fits, water leaks and occasional engine and gearbox faults. Things like that did not help the British hold their market and when American customers, who were buying imports anyway, saw the Datsun gaining in popularity, those who had not tried them before did now.

The MGB GT V-8 was a slab of a car compared with the stylish 260Z – as was the TR6 – though it was a substantial improvement on the performance of the 1,800cc engined MGB and stood head and shoulders above

Facing the Competition

The Datsun 260Z and its Adversaries, 1974

Car and Model	Engine Type and Size	Power (bhp)	Gearbox	Max. Speed (kph)	Max. Speed (mph)	0–60mph (secs)	GB Price
Datsun 260Z	6 cyl in line water cooled OHC 2,565cc	162	4-speed manual	204	127	10.0	£2,895
Alfa Romeo Montreal	V-8 water cooled OHC 2,593cc	200	5-speed manual	225	140	7.6	£6,250
Ferrari Dino 246GT	V-6 water cooled DOHC 2,418cc	175	5-speed manual	227	141	7.9	£6,246
Jensen Healey	4 cyl in line water cooled DOHC 1,973cc	140	4-speed manual	193	120	9.7	£2,190
Lotus +2S	4 cyl in line water cooled DOHC 1,558cc	130	4-speed manual	193	120	7.5	£2,793
Maserati Merak	V-6 water cooled OHC 2,965cc	190	5-speed manual	238	148	8.2	£7,000
Porsche 911S	Flat-6 air cooled OHC 2,341cc	190	5-speed manual	233	145	8.1	£5,943
Triumph TR7	4 cyl in line water cooled OHC 1,998cc	90	4-speed manual	174	108	11.3	£3,146
TVR 3000M	V-6 water cooled OHV 2,994cc	142	4-speed manual	201	125	7.7	£2,484

Facing the Competition

the earlier MGC in balance, handling, even comfort and certainly top speed and acceleration. The MGB GT V-8 could reach 60mph from a standstill in 8.6 seconds, while its top speed was a close match for the 260Z. In Britain there was a tuned version of the V-8 MG called the Costello V-8 MGB GT, which improved the top speed by 5mph (8kph) and the acceleration to 60mph to 7.8 seconds. In the United States the MG V-8 had its loyal followers in the hard-core MG fraternity, but its market was never going to be large.

The TVR 3000M was a sharp car on the road, just like its predecessor, but was now endowed with a bigger engine that gave it a 124mph (200kph) top speed (but a slower 0–60mph time of 7.7 seconds). Ford's Capri 2800 showed no performance improvement on the smaller-engined 2600 that preceded it, though it was beginning to show itself as a car that appealed to a less sporting sector of the market, so it was not a real competitor to the 260Z. But another German product was – the Porsche 924, new in 1975 and catching on very quickly both in Europe and in the United States.

The Porsche 924 was a front-engined rear drive car, the first ever such car from Zuffenhausen. The gearbox and final drive were combined in a transaxle to provide as near a 50/50 front/rear balance as possible and its handling was superb. It was a car of stunning looks, too, for its time – so much so that the 968 of the 1990s is essentially an updated 924. Its 2-litre water-cooled engine produced 125bhp and its four-speed gearbox helped it to 125mph (201kph), though it took 11.9 seconds to reach 60mph from a standstill. It was a two-plus-occasional-two and set the Porsche enthusiast world buzzing, costing over $4,000 less than the 911 2.7 at $9,400, but around $3,000 more than the 260Z. The 2.7-litre 911 Carrera, on the other hand, was also a stunning handler and could turn in 144mph (232kph), reaching 60 mph in only 7.5 seconds – but it was priced at over $13,500.

In Milan and Modena, the Italian manufacturers were producing some pretty exciting cars. Against the 260Z models, Alfa Romeo were fielding the Alfetta GTV 2000, a four-seater very cleverly designed by Giorgetto

Porsche's 924 was Zuffenhausen's first front-engined car and stood for some time as the target market model for Nissan with the 'Z' cars.

Facing the Competition

The Datsun 260Z 2+2 and its Adversaries, 1974–75

Car and Model	Engine Type and Size	Power (bhp)	Gearbox	Max. Speed (kph)	Max. Speed (mph)	0–60mph (secs)	GB Price
Datsun 260Z 2+2	6 cyl in line water cooled OHC 2,565cc	162	4-speed manual	204	127	10.0	£3,125
Alfa Romeo 2000GTV	4 cyl in line water cooled DOHC 1,962cc	131	4-speed manual	187	116	9.4	£2,849
Ferrari Dino 308GT4	V-8 water cooled DOHC 2,924cc	242	5-speed manual	245	152	6.4	£8,340
Lamborghini Urraco	V-8 water cooled OHC 2,463cc	220	5-speed manual	238	148	8.5	£7,567
Lotus Elite 501	4 cyl in line water cooled DOHC 1,973cc	155	5-speed manual	206	128	7.8	£5,751
Porsche 911S	Flat-6 air cooled OHC 2,341cc	190	5-speed manual	233	145	8.1	£5,943
Reliant Scimitar GTE	V-6 water cooled OHV 2,994cc	135	4-speed manual	196	122	10.0	£3,042
Triumph Stag	V-8 water cooled OHC 2,997cc	146	4-speed + overdrive	193	120	10.4	£2,436

Facing the Competition

The Alfetta GTV 2000 (above) and the Ferrari 308GTB (below) were Northern Italy's challenge to the 'Z' car, but both were very expensive by comparison.

Giugiaro that looked much like a 2+2, handled very well and came in with a 121mph (195kph) maximum. The driving position was long in the arm and short in the leg, but it was quite an adversary, though expensive against the Datsun 260Z in either the two-seat or 2+2 form. On the other hand, Ferrari now had either the 308 GTB or GT4. With the V-8 3 litre engine these Ferraris were now lifted above the Datsuns, as they were 155mph (249kph) projectiles capable of 60mph in under 7 seconds from 250bhp.

Facing the Competition

The Datsun 280Z and its Adversaries, 1975

Car and Model	Engine Type and Size	Power (bhp)	Gearbox	Max. Speed (kph)	Max. Speed (mph)	0–60mph (secs)	GB Price
Datsun 280Z	6 cyl in line water cooled OHC 2,753cc	162	4- or 5-speed manual	204	127	10.0	£3,900*
AC 3000ME	V-6 water cooled OHV 2,994cc	140	5-speed manual	217	135	8.0	£6,000*
Ferrari 308GTB	V-8 water cooled DOHC 2,926cc	255	5-speed manual	253	157	6.5	£11,251
Lotus Esprit	4 cyl in line water cooled DOHC 1,973cc	155	5-speed manual	222	138	8.4	£7,883
Porsche 924	4 cyl in line water cooled OHC 1,984cc	125	5-speed manual	201	125	11.1	£7,350
Triumph TR7	4 cyl in line water cooled OHC 1,998cc	90	4-speed manual	174	108	11.3	£3,146
TVR 3000M	V-6 water cooled OHV 2,994cc	142	4-speed manual	201	125	7.7	£3,990

* Estimate, converted to sterling for comparison purposes and allowing for import duties and taxes

Facing the Competition

The Datsun 280Z 2+2 and its Adversaries, 1976

Car and Model	Engine Type and Size	Power (bhp)	Gearbox	Max. Speed (kph)	Max. Speed (mph)	0–60 mph (secs)	GB Price
Datsun 280Z 2+2	6 cyl in line water cooled OHC 2,573cc	162	4-speed manual or automatic	204	127	11.0	£4,125*
Alfa Romeo Alfetta GTV 2000	4 cyl in line water cooled DOHC 1,962cc	121	5-speed manual	193	120	8.9	£4,779
BMW 633 CSi	6 cyl in line water cooled OHC 3,210cc	200	4-speed manual	209	130	9.7	£13,599
Ferrari Dino 308GT4	V-8 water cooled DOHC 2,924cc	242	5-speed manual	245	152	6.4	£11,450
Lotus Elite 503	4 cyl in line water cooled DOHC 1,973cc	155	5-speed manual	196	122	7.8	£8,357
Porsche 911 2.7	Flat-6 air cooled OHC 2,687cc	167	5-speed manual	222	138	7.5	£12,100
Reliant Scimitar GTE	V-6 water cooled OHV 2,994cc	135	4-speed manual	196	122	10.0	£4,907
Triumph Stag	V-8 water cooled OHC 2,997cc	146	4-speed + overdrive	193	120	9.8	£5,177

* Estimate, converted to sterling from US$, for comparison purposes and allowing for import duties and taxes

Facing the Competition

Meanwhile, back in Britain, another result of Giorgetto Giugiaro's design work went into production in 1975. Built in Norfolk, it was the Lotus Esprit, a sleek glass-fibre bodied two-seater powered by a 2 litre twin cam engine developed by Lotus from a Vauxhall engine. Even with such a small engine, the Lotus gave out 155bhp and via its five-speed gearbox it reached 135mph (217kph). Despite its price, it was a true competitor for the 260Z. Lotus had to work at their sales, but they had overcome much of the reputation for unreliability that had plagued them with the original Elite in the 1960s and their Grand Prix racing record was a great help to their commercial success.

GROWING UP WITH THE 280Z AND 280ZX

The 280Z two-seater was introduced into the United States in 1975 and was followed by the worldwide introduction of the 280ZX three years later. These were now more up-market cars – the price tag of the 280ZX in 1978 was only $2,500 behind the Porsche 924, a lot less than the gap had been with the 260Z. Legislation was playing its part in levelling out prices to a certain degree, because emission and structural strength rules were bringing cars closer to each other in very broad specification terms.

From Britain, the Lotus line-up, despite the smaller engine, was a superb match on performance for the 280Z. The Esprit two-seater had been revamped by 1979 and the Series 2 had styling revisions but, more important, better air feed to the twin Dellorto carburettors, which meant 5bhp more and improved acceleration, though with a price tag that was a clear $5,000 above the 280ZX. The other two Lotus cars were the Elite four-seater, with a slightly lower top speed than the Esprit, and an even higher price, while the Eclat 2+2 fell into the middle of the range. Ford, on the other hand, was well below the price tag of the 280ZX, had a car that held four people in similar comfort, came close on top speed, but fell far short on handling characteristics and roadholding.

The Triumph Stag had come and gone, but now the TR7 was on sale, a 2-litre two-seater offered in roadster and coupé forms. In its original 2-litre form, the TR7 was no match for the 2.8-litre Datsun, but when the Rover 3.5-litre engine was installed – the engine many said should have gone into the Stag – the TR8 came nearer to the 280Z, though its performance still fell short and its record of reliability seems to have left a little to be desired. On the other hand, the TVR Taimar hatchback was a real match for the Datsun two-seater. In naturally aspirated form the Taimar had a 0–60mph time of 7.8 seconds, while the Turbo version reached 145mph (233kph), and the price was in the same range.

Still, the only true German competitor for the 280ZX still came from Porsche, in the form of the 911, if you wanted to pay the price, and the 924, which was also now available in turbocharged form. The bottom-of-the-range 924 still produced 125mph (201kph) and now reached 60mph in 9.6 seconds, while the 924 Turbo would do 0–60mph in 7.2 seconds and ran up to 144mph (232kph). The 911 was now a 3 litre and still had a price almost double that of the 280ZX – but the performance was still electrifying, with a top speed over 140mph (225kph) and a 0–60mph time of only 6.7 seconds. And then there was the new 928, more than double the price of any Datsun, but capable of propelling four people in comfort at 145mph (233kph) where that speed could be legally reached.

If price was not the principal factor, Italian contenders came from Alfa Romeo, Ferrari and Maserati. Alfa Romeo had by 1981 introduced the GTV6 2.5 Alfetta Coupé.

Facing the Competition

The Lotus Esprit was another Giugaro styling exercise which was immensely successful and, despite its smaller engine, could outrun a 'Z' car, although it cost a lot more.

This was another 130mph (209kph) machine that took 8.8 seconds to reach that magic 60mph. A pretty comfortable four-seater, the GTV6 was also a fine handler and even with four people in it could be driven with quite a bit of enthusiasm. To a few other than the dedicated Alfa Romeo fans, this car was thought worth the extra cash for what it had to offer. It was not enough, though, to ruffle the feathers of Nissan, as the lost sales were not enough to worry about. Nor would it be so with either the Ferrari or Maserati – the Mondial and Merak being their offerings.

The Ferrari Mondial 8 was not quite the stunner that earlier small Ferraris had been, but offered seating for four, the rear two being no more or less cramped than in a 280ZX 2+2, with a top speed approaching 140mph (225kph). The 0–60mph time was

Alfa Romeo was finally to take up the challenge of the 280ZX by 1981, with the GTV-6-2.5 which was a real road-burning four seater.

The Datsun 280ZX and its Adversaries, 1979

Car and Model	Engine Type and Size	Power (bhp)	Gearbox	Max. Speed (kph)	Max. Speed (mph)	0–60mph (secs)	GB Price
Datsun 280ZX	6 cyl in line water cooled OHC 2,753cc	135	5-speed manual	204	127	9.2	£8,629
AC 3000ME	V-6 water cooled OHV 2,994cc	140	5-speed manual	217	135	8.0	£10,500*
Ferrari 308GTB	V-8 water cooled DOHC 2,926cc	255	5-speed manual	243	151	7.9	£18,973
Lotus Esprit S2	4 cyl in line water cooled DOHC 1,973cc	160	5-speed manual	217	135	8.0	£14,176
Porsche 924	4 cyl in line water cooled OHC 1,984cc	125	5-speed manual	201	125	11.1	£7,350
Triumph TR8	V-8 water cooled OHV 3,528cc	137	5-speed manual	193	120	8.4	£6,400*
TVR Taimar	V-6 water cooled OHV 2,994cc	142	4-speed manual	201	125	7.7	£9,859

* Estimate

The Datsun 280ZX 2+2 and its Adversaries, 1979

Car and Model	Engine Type and Size	Power (bhp)	Gearbox	Max. Speed (kph)	Max. Speed (mph)	0–60 mph (secs)	GB Price
Datsun 280ZX 2+2	6 cyl in line water cooled OHC 2,753cc	135	5-speed manual or automatic	204	127	11.2	£8,660
Alfa Romeo Alfetta GTV 2000	4 cyl in line water cooled DOHC 1,962cc	121	5-speed manual	193	120	8.9	£6,020
BMW 635CSi	6 cyl in line water cooled OHC 3,453cc	218	5-speed manual	225	140	8.5	£18,739
Ferrari Dino 308GT4	V-8 water cooled DOHC 2,926cc	205	5-speed manual	222	138	7.8	£15,450
Lotus Elite 503	4 cyl in line water cooled DOHC 1,973cc	160	5-speed manual	201	125	9.7	£14,676
Mercedes Benz 450SLC	V-8 water cooled OHV 4,520cc	225	3-speed automatic	201	125	9.3	£20,987
Porsche 924	4 cyl in line water cooled OHC 1,984cc	125	5-speed manual	201	125	11.0	£9,103
Reliant Scimitar GTE	V-6 water cooled OHV 2,994cc	135	4-speed manual	196	122	10.0	£8,137
Toyota Celica Supra	6 cyl in line water cooled OHC 2,563cc	113	5-speed manual	180	112	10.2	£6,463

The Datsun 280ZX Turbo and its Adversaries, 1982

Car and Model	Engine Type and Size	Power (bhp)	Gearbox	Max. Speed (kph)	Max. Speed (mph)	0–60mph (secs)	US Price
Datsun 280ZX Turbo	6 cyl in line water cooled turbocharged OHC 2,753cc	180	5-speed manual	208	129	7.4	$18,299
De Lorean Coupé	V-6 water cooled OHC 2,849cc	130	5-speed manual	177	110	10.5	$25,600
Ferrari 308GTB	V-8 water cooled DOHC 2,926cc	255	5-speed manual	253	157	6.5	$56,500
Lotus Esprit Turbo	4 cyl in line water cooled turbocharged DOHC 2,174cc	210	5-speed manual	245	152	6.1	$25,500
Porsche 924 Turbo	4 cyl in line water cooled turbocharged OHC 1,984cc	154	5-speed manual	204	127	9.2	$21,500
Mazda RX-7	Twin rotor water cooled Rotary	118	5-speed manual	190	118	9.7	$11,895
Chevrolet Corvette	V-8 water cooled OHV 5,735cc	190	4-speed manual	201	125	9.2	$22,537

Ferrari's Mondial 8 was Modena's response to the Nippon invader.

surprisingly slow, at 9.9 seconds – slower than for any Ferrari hitherto, but you could eventually outrun the Japanese car if you had enough distance. And of course there were the charisma and handling of Ferrari – at a price more than two and a half times that of the Datsun. The Maserati Merak, a two-seater, was only twice the price of a 280ZX. Its 3-litre engine took it to over 150mph (241kph), and it scratched 60mph at 7.7 seconds.

The Lotus 2.2 litre Esprit Turbo was now available, and this was giving 0–60mph in a stunning 6.1 seconds, with a top speed to match the Maserati, for £2,000 less in Britain. The Italian contenders for a chunk of the Datsun market were really not in contention, for Alfa Romeo, Ferrari and Maserati all had their own followings and took very little from Datsun's market.

CONTENDERS FOR THE CROWN OF THE 300ZX

The original 300ZX was only a stepping stone to where Nissan was aiming to be in the sports car market, for it was ultimately to become Japan's first true Supercar. It now had to compete with Supercars in order to take its place – indeed the Nissan 300ZX in its final form was to attack the upper sector of the car market in the same way as its now illustrious predecessor had done in a lower price sector twenty years before. But now there was serious competition from the home market too.

At home, Nissan had to face the Toyota Supra, a 3-litre sports car with a fuel injected six cylinder engine that delivered 135mph (217kph) and made 60mph in just 8 seconds. In 1986/87 the Supra was styled remarkably like the 300ZX and from a rear quarter view the casual observer would have to take a second look to be sure what it was. The pricing of the two cars was also remarkably close, though in Britain the Toyota cost about £200 more. While it was certainly the fastest production car Toyota had ever produced, the Nissan still gave it 7mph (11kph) and over a second on the 0–60mph time.

Mitsubishi produced its flagship Starion as a 2-litre turbocharged car. Styling can best be described as possessed of a little individuality mixed with quite a bit of the lines from Nissan and Toyota. Its four-cylinder engine produced 177bhp and propelled

Facing the Competition

The Nissan 300ZX and its Adversaries, 1984–85

Car and Model	Engine Type and Size	Power (bhp)	Gearbox	Max. Speed (kph)	Max. Speed (mph)	0–60mph (secs)	US Price
Nissan 300ZX	V-6 water cooled OHC turbocharged 2,960cc	200	5-speed manual	214	133	7.4	$19,699
Mitsubishi Starion	4 cylinder in line water cooled turbocharged 2,555cc	145	5-speed manual	193	120	9.2	$14,489
Ferrari 308GTB	V-8 water cooled DOHC 2,926cc	230	5-speed manual	229	142	6.8	$53,125
Lotus Turbo Esprit	4 cyl in line water cooled DOHC turbocharged 2,174cc	205	5-speed manual	238	148	6.6	$49,500
Porsche 944	4 cyl in line water cooled OHC 2,479cc	143	5-speed manual	201	125	9.0	$22,300
Lamborghini Jalpa	V-8 water cooled DOHC 3,485cc	250	5-speed manual	214	133	7.3	$53,125

Facing the Competition

The Nissan 300ZX 2+2 and its Adversaries, 1984–85

Car and Model	Engine Type and Size	Power (bhp)	Gearbox	Max. Speed (kph)	Max. Speed (mph)	0–60mph (secs)	US Price
Nissan 300ZX 2+2	V-6 water cooled OHC turbocharged 2,960cc	200	5-speed manual	214	133	7.4	$20,230
Alfa Romeo Alfetta GTV6	V-6 water cooled DOHC 2,492cc	160	5-speed manual	209	130	8.2	$19,500
BMW 635CSi	6 cyl in line water cooled OHC 3,430cc	182	5-speed manual	212	132	8.0	$40,705
Ford SVO Mustang	4 cyl in line water cooled OHC turbocharged 2,300cc	175	5-speed manual	216	134	7.7	$14,521
Porsche 944	4 cyl in line water cooled OHC 2,479cc	143	5-speed manual	201	125	9.0	$22,300
Toyota Supra	6 cyl in line water cooled OHC 2,759cc	161	5-speed manual	204	127	8.4	$16,558

the car to 143mph (230kph), reaching 60mph after just 6.5 seconds. Falling just below the Nissan on price, the Mitsubishi was quite a well made car with a high specification, aimed clearly at the 'executive' market that, in Britain at least, was expressing some interest in semi-exotic cars.

Mazda, which had begun its entry into the motor industry by making three wheeled utility vehicles, was probably the most adventurous Japanese car maker in its engineering innovation. This was the manufacturer that took on Dr Felix Wankel's rotary engine and made it work reliably and efficiently. That engine concept, offered in nominally 2.4-litre twin-rotor form in a coupé called the RX-7, brought 135mph (217kph) and 0–60mph in 8.5 seconds to the 'executive' market in Britain for just under £15,000. This 2+2 was not so much aimed at taking on the Nissan 300ZX as at the Porsche 924, offering a similar performance and appointment level for about £3,000 less.

By 1989 the new Nissan 300ZX was in a different league again. A quite deliberate policy decision took the car into the true Supercar league, where it now had to contend for market share, in a much smaller market, with such machinery as the Aston Martin DB7, Ferrari 348, Mercedes Benz 500SEC, Porsche 944 – and perhaps it came closer to competing with the magical and elusive 911 than ever before. Most of these cars were priced well above the 300ZX, though the 944, which had been a performance and specification target model for Nissan since the inception of their newest, most elegant sports car, was priced in the same bracket that was aimed for by Nissan.

It was a hard lesson for Nissan to learn that the European sports car market at its upper end had certain loyalties that would be hard to break. It was made even more difficult to penetrate by British legislation proposals on emission standards that would demand an extra catalyst, which would have meant expensive re-design work on a car that was seen as perfectly satisfactory in its main market, the United States. Nissan realized that they were never going to achieve optimum sales targets in Britain; deciding that the market was not big enough to sustain the support essential to any high-profile car, they decided to withdraw the 300ZX in 1994 – just twenty five years after the first 240Z took the sports car world by storm.

Toyota couldn't, with the 3-litre Supra, take on the 'Z' cars in a straight fight, as they were short on both top speed and 0–60mph time.

Facing the Competition

The Nissan 300ZX 2+2 (Z-32) and its Adversaries, 1994

Car and Model	Engine Type and Size	Power (bhp)	Gearbox	Max. Speed (kph)	Max. Speed (mph)	0–60mph (secs)	GB Price
Nissan 300ZX 2+2	V-6 water cooled OHC twin-turbocharged 2,960cc	300	5-speed manual	249	155	5.9	£35,115
Aston Martin DB-7	6 in line water cooled supercharged DOHC 3,228cc	335	5-speed manual	266	165	5.7	£78,500
BMW 850CSi	V-12 water cooled OHC 5,576cc	296	5-speed manual	249	155	6.0	£79,750
Ferrari Mondial	V-8 water cooled DOHC 3,405cc	300	5-speed manual	254	158	6.3	£67,171
Mercedes Benz 500S Coupé	V-8 water cooled OHC 4,973cc	218	3-speed automatic	230	143	7.3	£74,600
Porsche 968 Turbo	4 cyl in line water cooled turbocharged OHC 2,990cc	220	5 speed manual	253	157	6.6	£40,695

10 Road Tests and Press Reviews

Car Life was one of the first magazines to review the Datsun 240Z after its announcement in late 1969. That was in February 1970, before anyone had been given the opportunity to road test the car properly. It was known about in Britain on its announcement, of course, but the public was not going to clap its eyes on the new Japanese threat to European sports cars for another year. The *Car Life* quotation about the lights going on all over Europe may not have been literally true, its inference certainly was – here was a car that was going to shake many a European car maker to their roots, for it was destined to take a large slice of their traditional market in North America away from them. It also established in the minds of middle-income America that the Japanese could build high quality, high specification cars that sold for a lot less than Detroit iron and performed at least as well. It was a watershed for the American private car market and the Datsun 240Z did as much to punch that home for Japan's motor industry as anything else.

OUT ON THE ROAD WITH THE 240Z

Sports Car Graphic did it first – a comparison test between the 240Z and two of its anticipated rivals. The rivals were the MGB GT and the Opel GT. It is hardly surprising that the 240Z should win, but it was a little unusual for the magazine's road tester to announce his verdict in the first sentence. He apologized for that, but felt such enthusiasm for the new Japanese car (this was March 1970 and the ink of the customs officer's rubber stamp on the shipping slip would barely have been dry) that he simply could not, on his own admission, contain himself.

The reviewer's comparisons were based on what he described as the cars in the most active GT sales market. The Opel was the application of the 'kitcar' concept, in that it was a new body on a 'nothing chassis', while the MG was seen as just a roadster with a new roof. The Datsun 240Z, on the other hand, was a completely new car from the ground up (remembering that it borrowed just a few bits from the existing parts bins). All three cars were put through their paces on the road and on a skidpan. Interestingly, the MG came out best on the skidpan, holding its line for longer than either of the other two. The author is pointed about the archaic engineering of the MG, but respectful about the car's handling in general, as it was still endowed with Syd Enever's 'touch', the last car to be so. The Opel, on the other hand, was quieter than the MGB, a reasonably comfortable and stylish car that would propel its occupants to 193kph (120mph), but which would roll them about all over the place when punished.

The Datsun, then, had the best road speed, the second best handling, though with a tendency to wander at high speed, the highest

Road Tests and Press Reviews

price (by a few dollars only) and by far the most promise of all three. It had the advantage of having been designed as a new project, so there were no preconceptions about how far ahead of its predecessor it should or should not go. There were less inhibitions about engineering or styling specifications, because there was nothing to alter. The new Datsun was to be a total departure from the SR311 Series, whereas the MGB used a lot of MGA inside, though the slab-like styling was about as far as you could get from the smooth and elegant lines of the MGA.

Road and Track put the 240Z through its paces and published the results in April 1970. The opening comments observed that Datsun had held a sizeable portion of the United States sports car market with the

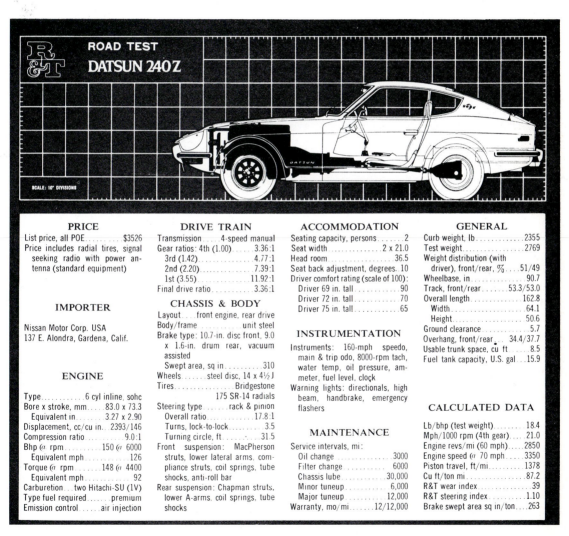

Road & Track *road test data panel, April 1970 (continued overleaf).*

Road Tests and Press Reviews

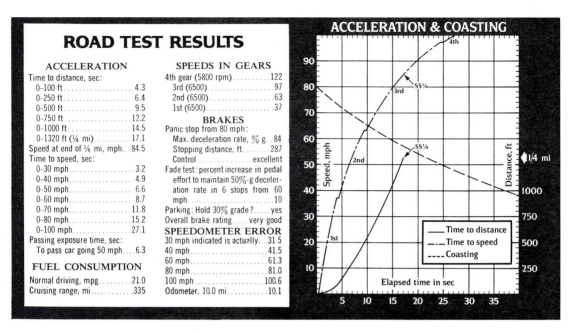

Road & Track *road test data panel, April 1970 (continued from previous page).*

SR311 1600 and its 2-litre successor. The reviewer went on to note that these two Datsuns were based on British sports car design influences – logical, they said, since the British were dominant in the market when the Datsuns arrived on the market. But they represented the old order of stark, open two-seaters that had strong, noisy, engines and few concessions to ride comfort, performance being the most important factor.

The review went on then to remind us that the 240Z was the result of a very careful research programme, including studies of the European design practices, in Germany and Italy for example. The result was the extremely stylish and thoroughly modern 240Z. *Road and Track*'s view of the design lines, the interior layout and trim, the engineering of the Z, was that it was superb and so completely well thought out that this car was destined for great things. The reviewer was impressed with the handling and performance and was very taken with the detailed thinking that brought such clever little features as the fuse-board set in behind the ashtray.

With the observation that it was quiet and smooth at low engine speeds, it seemed that an unpleasant rumble developed as the car was wound up and it appeared to peak a little below the quoted peak engine speed of 6,000rpm, at around 5,500 in fact. The steering was found to be sensitive to minor adjustments and fuel economy on the test car was found to be poor, though it was thought not to be characteristic. The ride was thought to be good, but a little erratic on rough surfaces. A rear axle whine manifested itself at over 50mph, becoming worse up to 70mph (113kph) and then disappearing. In general, it was thought that the Datsun 240Z was destined to be a major sales success – a 'super bargain' is how its $3,526 price tag was described. There were a few rough edges to be rounded off, like keeping water out of the brakes and dealing with that rumble and wander, but it was destined in the opinion of *Road and Track* to be a best-seller, which it surely was.

RIDING THE 260Z

A lot of the lessons learned from the 240Z, like improving the shock absorbers to give a less bumpy ride on indifferent surfaces, improving the water-shedding capability of the rear brake drums and keeping noises out of the passenger compartment, were incorporated into its successor. *Car and Driver* thought the 260Z had much better handling, saying it was so stable that normal interruptions of sharp increases and decreases of throttle, braking and wet surface handling had little effect. The road tester's view was that the 260Z was the kind of car that persuaded the driver to avoid freeways and search out the twisty back roads to enjoy the drive! John Bolster, writing in Britain's *Autosport*, thought the 260Z to be a charming road car for any purpose, with plenty of performance in hand and a sporting character that was not over-emphasized.

Road and Track opened their road test of the 260Z with the comment 'They have done all the right things to the legendary bargain GT' and continued:

> Rarely do we mark the passing of a car that we have in the recent past rated as one of the world's most important Sports-GT cars with a more blase attitude than the Datsun 240Z. When the Z-car was introduced four years ago it revolutionised the Sports-GT market in America and was such an incredible buy that at that price, we tended to dismiss some of its failings, which were few anyway. But in the ensuing three years, the plunge of the dollar, the upward-valued yen and the near $5000 price tag dictated a more critical look. On close inspection, the 240Z was found wanting in two important areas: driveability and highspeed stability. So the 240Z is gone and R&T will not mourn its passing. But hold on, Z-Car lovers ... the 240Z is dead – but in its place is the 260Z, a car that goes a long way towards eliminating all the criticism we levelled at the car last year.

Of course, the 240Z had suffered the ravages of legislation, becoming heavier and getting choked up with emission controls, the result being that its power output was reduced, its 0–60mph time suffered badly and its handling was also degraded. The *Road and Track* reviewer was quick to pick up these points, but was equally quick to point out that, within the scope available, the 260Z went a long way to cure those problems. It is interesting here to compare the time of 0–60mph achieved by *Road and Track*, who were not renowned for being light-footed on road tests, and that set by *Motor* magazine in Britain – one car being emission controlled, the other not. *Road and Track* managed 10 seconds dead, whereas *Motor* cut 1.2 seconds from that. Even this showed that the 260Z engine had been deliberately de-tuned to reduce its output of noxious gases, though it had good torque for bottom-end performance.

Notwithstanding the de-tuning of the engine – and *Road and Track* reminds us of that by pointing out that the 260Z was not as fast as its predecessor – there were plus points according to the tester. Improved driveability was the most important: the tendency to wander at speed was virtually eliminated. Another reason for lower top speed was the reduction in peak engine revolutions, as the piston travel was increased with the longer stroke unit, so to hold piston speed, the rev limit had to be reduced. It was also noted that the longer and heavier front bumper helped to keep the nose

ACCELERATION standing ¼ mile, seconds

- DATSUN 260Z
- FIAT 124 SPIDER (1973)
- JENSEN-HEALEY (1973)
- PORSCHE 914 2.0 (1973)

BRAKING 70-0 mph, feet

- DATSUN 260Z
- FIAT 124 SPIDER (1973)
- JENSEN-HEALEY (1973)
- PORSCHE 914 2.0 (1973)

FUEL ECONOMY RANGE mpg

- DATSUN 260Z
- FIAT 124 SPIDER (1973) NOT AVAILABLE
- JENSEN-HEALEY (1973)
- PORSCHE 914 2.0 (1973)

PRICE AS TESTED dollars x 1000

- DATSUN 260Z
- FIAT 124 SPIDER (1973)
- JENSEN-HEALEY (1973)
- PORSCHE 914 2.0 (1973)

INTERIOR SOUND LEVEL dBA

(70 mph cruise / Full throttle acceleration)

- DATSUN 260Z
- FIAT 124 SPIDER (1973)
- JENSEN-HEALEY (1973)
- PORSCHE 914 2.0 (1973)

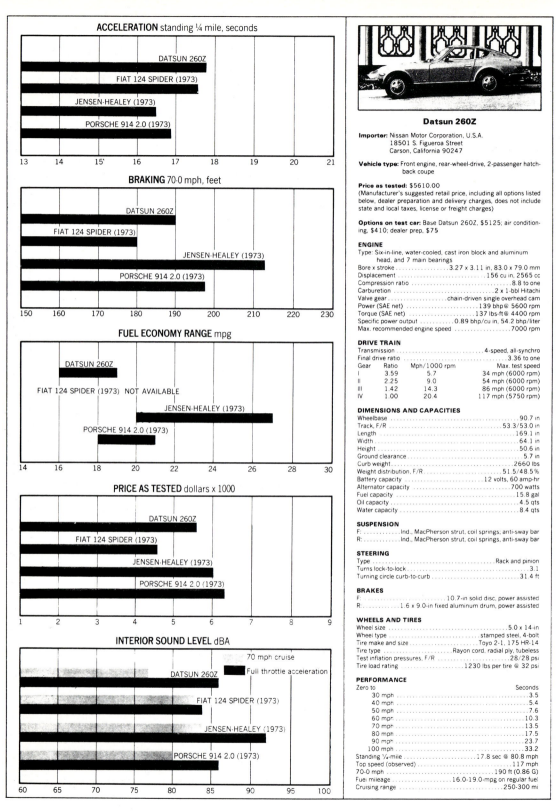

Datsun 260Z

Importer: Nissan Motor Corporation, U.S.A.
18501 S. Figueroa Street
Carson, California 90247

Vehicle type: Front engine, rear-wheel-drive, 2-passenger hatch-back coupe

Price as tested: $5610.00
(Manufacturer's suggested retail price, including all options listed below, dealer preparation and delivery charges, does not include state and local taxes, license or freight charges)

Options on test car: Base Datsun 260Z, $5125; air conditioning, $410; dealer prep, $75

ENGINE
Type: Six-in-line, water-cooled, cast iron block and aluminum head, and 7 main bearings
Bore x stroke . . . . 3.27 x 3.11 in, 83.0 x 79.0 mm
Displacement . . . . 156 cu in, 2565 cc
Compression ratio . . . . 8.8 to one
Carburetion . . . . 2 x 1-bbl Hitachi
Valve gear . . . . chain-driven single overhead cam
Power (SAE net) . . . . 139 bhp @ 5600 rpm
Torque (SAE net) . . . . 137 lbs-ft @ 4400 rpm
Specific power output . . . . 0.89 bhp/cu in, 54.2 bhp/liter
Max. recommended engine speed . . . . 7000 rpm

DRIVE TRAIN
Transmission . . . . 4-speed, all-synchro
Final drive ratio . . . . 3.36 to one

Gear	Ratio	Mph/1000 rpm	Max. test speed
I	3.59	5.7	34 mph (6000 rpm)
II	2.25	9.0	54 mph (6000 rpm)
III	1.42	14.3	86 mph (6000 rpm)
IV	1.00	20.4	117 mph (5750 rpm)

DIMENSIONS AND CAPACITIES
Wheelbase . . . . 90.7 in
Track, F/R . . . . 53.3/53.0 in
Length . . . . 169.1 in
Width . . . . 64.1 in
Height . . . . 50.6 in
Ground clearance . . . . 5.7 in
Curb weight . . . . 2660 lbs
Weight distribution, F/R 51.5/48.5%
Battery capacity . . . . 12 volts, 60 amp-hr
Alternator capacity . . . . 700 watts
Fuel capacity . . . . 15.8 gal
Oil capacity . . . . 4.5 qts
Water capacity . . . . 8.4 qts

SUSPENSION
F: . . . . Ind., MacPherson strut, coil springs; anti-sway bar
R: . . . . Ind., MacPherson strut, coil springs, anti-sway bar

STEERING
Type . . . . Rack and pinion
Turns lock-to-lock 3.1
Turning circle curb-to-curb . . . . 31.4 ft

BRAKES
F: . . . . 10.7-in solid disc, power assisted
R: . . . . 1.6 x 9.0-in fixed aluminum drum, power assisted

WHEELS AND TIRES
Wheel size . . . . 5.0 x 14-in
Wheel type . . . . stamped steel, 4-bolt
Tire make and size . . . . Toyo 2-1, 175 HR-14
Tire type . . . . Rayon cord, radial ply, tubeless
Test inflation pressures, F/R . . . . 28/28 psi
Tire load rating . . . . 1230 lbs per tire @ 32 psi

PERFORMANCE
Zero to   Seconds
30 mph 3.5
40 mph 5.4
50 mph 7.6
60 mph 10.3
70 mph 13.5
80 mph 17.5
90 mph 23.7
100 mph 33.2
Standing ¼-mile . . . . 17.8 sec @ 80.8 mph
Top speed (observed) . . . . 117 mph
70-0 mph . . . . 190 ft (0.86 G)
Fuel mileage . . . . 16.0-19.0-mpg on regular fuel
Cruising range . . . . 250-300 mi

Car & Driver test data on the 260Z.

R&T ROAD TEST
DATSUN 260Z

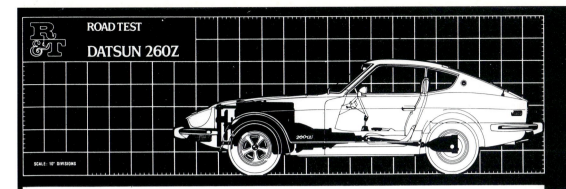

SCALE: 10" DIVISIONS

PRICE
List price, all POE $4995
Price as tested $5470
Price as tested includes standard equipment (AM/FM radio with electric antenna), air conditioning ($400), dealer prep ($75)

IMPORTER
Nissan Motor Corp in USA
18501 S Figueroa Ave
Carson, Calif. 90247

GENERAL
Curb weight, lb 2665
Test weight 2915
Weight distribution (with driver), front/rear, % 51/49
Wheelbase, in. 90.7
Track, front/rear 53.3/53.0
Length 169.1
Width 64.1
Height 50.6
Ground clearance 5.7
Overhang, front/rear 37.6/40.8
Usable trunk space, cu ft 8.5
Fuel capacity, U.S. gal. 15.8

ENGINE
Type .... sohc inline 6
Bore x stroke, mm .... 83.0 x 79.0
Equivalent in. .... 3.27 x 3.11
Displacement, cc/cu in. 2565/157
Compression ratio 8.8:1
Bhp @ rpm, net .... 139 @ 5200
Equivalent mph 116
Torque @ rpm, lb-ft .... 137 @ 4400
Equivalent mph 98
Carburetion .... two Hitachi-SU (1V)
Fuel requirement .... regular, 91-oct
Emissions, gram/mile:
 Hydrocarbons 2.7
 Carbon Monoxide 22.0
 Nitrogen Oxides 1.3

DRIVE TRAIN
Transmission .... 4-sp manual
Gear ratios: 4th (1.00) 3.36:1
 3rd (1.42) 4.77:1
 2nd (2.25) 7.56:1
 1st (3.59) 12.06:1
Final drive ratio 3.36:1

CHASSIS & BODY
Layout .... front engine/rear drive
Body/frame .... unit steel
Brake system .... 10.7-in. disc front, 9.0 x 1.6-in. drum rear, vacuum assisted
Swept area, sq in. 310
Wheels .... steel disc, 14x5
Tires .... Bridgestone 175HR-14
Steering type .... rack & pinion
 Overall ratio 18.0:1
 Turns, lock-to-lock 3.1
 Turning circle, ft 31.4
Front suspension .... MacPherson struts, lower lateral arms, compliance struts, coil springs, tube shocks, anti-roll bar
Rear suspension ..... Chapman struts, lower A-arms, coil springs, tube shocks, anti-roll bar

INSTRUMENTATION
Instruments: 160-mph speedo, 8000-rpm tach, 99,999 odo, 999.9 trip odo, oil press, coolant temp, ammeter, fuel level, clock
Warning lights: brake system, choke on, rear window heat, seatbelts, hazard, high beam, directionals

ACCOMMODATION
Seating capacity, persons 2
Seat width .... 2 x 21.0
Head room 36.5
Seat back adjustment, deg. 40

MAINTENANCE
Service intervals, mi:
 Oil change 3000
 Filter change 6000
 Chassis lube 30,000
 Minor tuneup 6000
 Major tuneup 12,000
Warranty, mo/mi 12/12,000

CALCULATED DATA
Lb/bhp (test weight) 21.0
Mph/1000 rpm (4th gear) 21.4
Engine revs/mi (60 mph) 2800
Piston travel, ft/mi 1450
R&T steering index 0.97
Brake swept area, sq in./ton 213

RELIABILITY
From R&T Owner Surveys the average number of trouble areas for all models surveyed is 12. As owners of earlier-model Datsun 240Zs reported 13 trouble areas, we expect the reliability of the 260Z to be average.

ROAD TEST RESULTS

ACCELERATION
Time to distance, sec:
 0-100 ft. 4.0
 0-500 ft. 9.7
 0-1320 ft (¼ mi) 17.9
Speed at end of ¼ mi, mph 78.5
Time to speed, sec:
 0-30 mph 3.7
 0-50 mph 7.6
 0-60 mph 10.0
 0-70 mph 13.9
 0-80 mph 18.7
 0-90 mph 24.7

SPEEDS IN GEARS
4th gear (5050) 113
3rd (6000) 96
2nd (6000) 60
1st (6000) 37

FUEL ECONOMY
Normal driving, mpg 20.0
Cruising range, mi (1 gal. res.) ... 296

HANDLING
Speed on 100-ft radius, mph .. 32.8
Lateral acceleration, g 0.720
Speed through 700-ft slalom, mph 51.3

BRAKES
Minimum stopping distances, ft:
 From 60 mph 165
 From 80 mph 272
Control in panic stop ...... excellent
Pedal effort for 0.5g stop, lb ... 33
Fade: percent increase in pedal effort to maintain 0.5g deceleration in 6 stops from 60 mph 30
Parking: hold 30% grade? ...... yes
Overall brake rating ...... very good

INTERIOR NOISE
All noise readings in dBA:
Idle in neutral 54
Maximum, 1st gear 86
Constant 30 mph 67
 50 mph 73
 70 mph 77
 90 mph 81

SPEEDOMETER ERROR
30 mph indicated is actually 28.0
50 mph 47.5
60 mph 57.0
70 mph 66.5
80 mph 75.5
Odometer, 10.0 mi 10.0

ACCELERATION
(graph showing Time to distance and Time to speed curves: 1st-2nd, 2nd-3rd, 3rd-4th, SS¼)

Road & Track's *test data for the 260Z in comparison with that of* Car & Driver *(opposite page).*

Road Tests and Press Reviews

The Aston Martin Lagonda, a four seater designed from the drawing board, which was shortened to become the DBS.

down (whether that was matter of the weight increase or the aerodynamic effect is not debated). The fact was that the impact-absorbing bumpers added 130lb (59kg) to the weight of the car and an extra 70lb (32kg) on the front end could well have had a beneficial effect on stability.

Interior improvements to the 260Z worthy of press comment included the revision of the interior door armrests and door pulls to the European-style integral type. *Road and Track* welcomed the demise of the plastic simulated wood-grain steering wheel, as it was slippery and felt unnatural, to be replaced with a simulated leather wheel, which was grippable and much more comfortable. Air conditioning was substantially improved, too, as the integral GM type installation was more efficient and easier to control. By and large, this reviewer found the 260Z a huge improvement over its predecessor and, even though the price had gone up on the original 1970 price tag of the 240Z, a sound explanation had been given for the excessive inflation to nearly $5,500 for the new model as tested.

That roadholding improvement mentioned by *Road and Track* was confirmed by *Car and Driver* and *Motor Trend*. Their views echoed *Road and Track* in their praise of the elimination of high speed wander. *Car and Driver* went on to comment that they found the 260Z so stable that variations in throttle or quite sharp braking and handling on wet surfaces seemed not to disturb the handling of the car. The 260Z certainly impressed all its testers as a worthy successor to the original Z and, interestingly, a 'retrospective' review of the 240Z and 260Z in *Classic and Sports Car* took the view that the 260Z was a great improvement on the 240Z and of the two, the 260Z was the car for the enthusiast to buy.

The 260Z 2+2 was road tested by *Road and Track* in May 1974. Not unreasonably, their staff thought the line of the 2+2 to be less pleasing than that of the two-seater, which brings to mind a comment made once by William Towns, the designer of Aston Martin's elegant DBS and V-8 saloons. He always averred that if the designer produced the line of the four-seater first, the two-seater would almost invariably look right and the line of the four-seater would be a much better result. The design that proved his point was the Aston Martin Lagonda, a four-door version of the Aston Martin DBS V-8, later just designated V-8, which to many was seen as a stretched version of the two-door car, whereas the two-door car was a shortened version of the original design. The 2+2 Datsun on the other hand was, like the Jaguar E-type, a stretch of the original two-seater that lost something in the stretching.

Despite the criticisms of the line, and there were clearly a lot of people who liked the car, the handling was found not to suffer. Indeed, the test driver in *Road and Track* found that a tendency of the two-seater to 'hop' had disappeared in the longer car, simply as a benefit of the longer wheelbase, which conferred a more comfortable ride. The tendency of the two-seater to jerk over large bumps had also gone and the general ride stability was found to be much improved. Part of the ride improvement can be put down to raised spring rates, though the extra weight of the car made it heavy to park. That said, it was also commented that steering out on the road was found to be reasonably light and quite precise, with a little understeer in hard driving. All in all, *Road and Track* was of the opinion that the decision to produce a 2+2 version of the Z was good, although it did criticize the amount of room in the back for the occasional rear seat passengers, commenting that only children or small adults would be reasonably comfortable.

LAST OF THE ORIGINAL LINE – THE 280Z

Because of the continuing pressures of emission legislation in the United States and because carburettors were becoming less practical, most manufacturers of cars aimed at the United States market were looking closely at fuel injection. Nissan were no exception, though almost everyone was also ready to accept that the only way to increase the power of 'clean' engines was to increase their size. So was born the Datsun 280Z, a fuel injected 2.8 litre version of the original 240Z.

Road and Track opens its initial review of the 280Z with the comment 'As the emission limits go down Datsun engine displacements go up' – verifying the point just made. *Road and Track* also makes the point in its June 1975 test of the 280Z that the general advantages of fuel injection are increased performance, superior cold starting, better driveability, resistance to vapour lock, improved fuel economy and better emission control. But the Datsun 280Z, according to their tester, was not about to confer upon its driver much in the way of improved performance or economy, because the main task undertaken by the manufacturer was the recovery of the old zest of the 240Z.

In general terms, the 'old zest' was recovered, but at a cost, both in cash price and in the weight of the new car. And just to unnerve the Californian buyer of a 280Z, there were two extra warning lights on the dashboard – the first to warn of the catalytic converter overheating and the other to tell the unsuspecting occupants that the floor of the car was overheating! It is small wonder, then, that there was a warning printed on a sticker on the sun visor reminding the driver not to park on grass, weeds or brush, as it could start a fire! The apprehension was fortunately confined to California, as buyers elsewhere in the United States were spared the catalyser and the problems that went with it.

Apart from the 'step' in spark advance at around 2,700rpm mentioned earlier, *Road and Track* seemed generally quite pleased with the refinements brought to the 280Z and aired the opinion that it had to be quite a car to provide the performance improvements over the 260Z. The fuel consumption still compared reasonably well with the old 240Z – and on 91 octane fuel. On the down side, the criticism of the new model's steering would have been eliminated with power assistance, but it was not there so parking remained a chore. Also, despite the jerkiness of the suspension having disappeared,

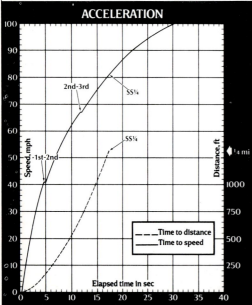

Road & Track's *test data on the Datsun 280Z two-seater.*

PRICE	
List price, all POE	$6284
Price as tested	$6784
GENERAL	
Curb weight, lb	2875
Weight distribution (with driver), front/rear, %	50/50
Wheelbase, in.	90.7
Track, front/rear	53.3/53.0
Length	173.2
Width	64.2
Height	51.0
Fuel capacity, U.S. gal.	17.2
CHASSIS & BODY	
Body/frame	unit steel
Brake system	10.7-in. discs front, 9.0 x 1.6-in. drums rear; vacuum assisted
Wheels	steel disc, 14 x 5J
Tires	Bridgestone Wide 70, 195/70-14
Steering type	rack & pinion
Turns, lock-to-lock	3.1
Suspension, front/rear	MacPherson struts, lateral arms, coil springs, tube shocks, a-r bar/Chapman struts, A-arms, coil springs, tube shocks, a-r bar
ENGINE & DRIVETRAIN	
Type	sohc inline 6
Bore x stroke, mm	86.1 x 79.0
Displacement, cc/cu in.	2754/168
Compression ratio	8.3:1
Bhp @ rpm, net	149 @ 5600
Torque @ rpm, lb-ft.	163 @ 4400
Fuel requirement	regular, 91-oct
Transmission	4-sp manual
Gear ratios: 4th (1.00)	3.55:1
3rd (1.31)	4.55:1
2nd (2.08)	7.38:1
1st (3.32)	11.79:1
Final drive ratio	3.55:1
CALCULATED DATA	
Lb/bhp (test weight)	21.2
Mph/1000 rpm (4th gear)	20.3
Engine revs/mi (60 mph)	2960
R&T steering index	1.08
Brake swept area, sq in./ton	196

ROAD TEST RESULTS	
ACCELERATION	
Time to distance, sec:	
0-100 ft	3.8
0-500 ft	9.6
0-1320 ft (¼ mi)	17.3
Speed at end of ¼ mi, mph	81.0
Time to speed, sec:	
0-30 mph	3.2
0-50 mph	6.9
0-60 mph	9.4
0-80 mph	16.8
0-100 mph	30.2
SPEEDS IN GEARS	
4th gear (5700 rpm)	119
3rd (6500)	104
2nd (6500)	67
1st (6500)	41
FUEL ECONOMY	
Normal driving, mpg	19.5
BRAKES	
Minimum stopping distances, ft:	
From 60 mph	130
From 80 mph	221
Control in panic stop	excellent
Pedal effort for 0.5g stop, lb	45
Fade: percent increase in pedal effort to maintain 0.5g deceleration in 6 stops from 60 mph	see text
Overall brake rating	very good
HANDLING	
Speed on 100-ft radius, mph	32.8
Lateral acceleration, g	0.720
Speed thru 700-ft slalom, mph	53.9
INTERIOR NOISE	
All noise readings in dBA:	
Constant 30 mph	68
50 mph	71
70 mph	75
SPEEDOMETER ERROR	
30 mph indicated is actually	28.5
60 mph	58.0
70 mph	67.5

some of the old high speed wander was detected and the brakes were now thought to be close to their limit, the 280Z retaining discs at the front and drums at the rear.

However, the 280Z was reckoned to be one of the best cars around for easy-to-reach, easy-to-use controls and switches. The air conditioning, new in 1974, was reckoned to be the best yet fitted to a Datsun, despite the reservation about the demisting quoted earlier, and the conclusion of the test team was that this was the most refined car in the series to date. The price was a cause for concern, though the tester was at pains to point out that at $6,300 the car was no longer a bargain, until the reader realized that the hike in price was not really Nissan's fault, as they had no control over fluctuations in exchange rates, nor could they avoid adding the costs of the consequences of ever-stricter emission and impact regulations being imposed year by year. In the context of prices of the time, *Road and Track* concluded that the 280Z was still a good buy overall.

THE ZX CHANGES THE SHAPE OF THE Z

The year 1978 brought a whole new perspective to the Z series. The weight penalty of tightening legislation had really forced Nissan to look again at its sports car model and produce something that would enable them to stay in the market competitively. The decision to take the Datsun up-market was quite deliberate, for Nissan was now looking hard at the upper end of the sports car market, believing that it could probably bring top-of-the-market performance to the less affluent sports car enthusiast – less

Road Tests and Press Reviews

The Ferrari Daytona (this one is a racer), said to be the model inspiration for the 280ZX.

affluent, that is, in terms of the person who wanted Porsche, Jaguar or smaller Ferrari performance – but just could not stump up the price tag. The 280ZX was reckoned to be the Datsun answer to that enthusiast.

Patrick Bedard, reviewing the new 280ZX for *Car and Driver* magazine, described the line of the new two-seater as 'having the suggestion of a Ferrari Daytona', justifying his point that the new car was wider, longer and lower than its predecessor. The reviewer observed that the market had swung strongly towards the Giugiaro style of razor-edged long swept lines, as shown in the Lotus Esprit, the Alfa Romeo GTV – perhaps more so the smaller engined Sprint, born out of the Alfasud, though not a contender for the ZX market – but went on to comment that the Datsun had been cleverly updated, yet retained some resemblance to its forebears. Mechanically, the new car was accurately described as the logical successor to the 280Z, though with a little less power, but much improved by the use of the previously optional five-speed gearbox positioned behind it.

The total revision of the suspension brought a combination of favourable and critical comment. Bedard felt that the semi-trailing arm rear suspension offered few benefits to the car's occupants, as handling could be adversely affected by the changing geometry, giving more camber and toe change than single struts, so offering handling changes under various conditions. While the ride was felt to be more akin to a luxury car than a sports car, the minor amount of self-steering over bumps and during hard braking was not to the car's credit. On the plus side, though, the higher geared steering – 2.7 turns lock to lock replacing the earlier rack that gave 3.5, aided by power assistance – meant there was a greater sense of sure-footedness under most driving conditions.

Bolstering the move into a higher standard of luxury was the equipment list for the interior of the car. The gadgetry catalogue included an array of lights to report that you were short of water in the screen washer, among other things, a four-speaker stereo/radio, electric door mirrors, a dual fuel gauge and even cruise control. Then there

Road Tests and Press Reviews

was the dual button electric door window on the driver's side, which would lower only as far as you wanted on one button, but a touch of the other would bring the window zooming down all the way, allowing you to keep both hands on the steering wheel more of the time. All these features were part of the 'Grand Luxe' package, which also brought you velour upholstered seats. The ZX was now set to pitch for the market held by Porsche with the 924, by Alfa Romeo's GTV6 2.5 and, of course, Toyota's Celica Supra in the United States, to say nothing of the domestic products it would displace.

Moving on to the 2+2, *Car and Driver* took the view that the line of the new occasional four-seater was softer and more appealing than the immediately previous model and, with the manufacturer 'turning up' the comfort, this was a car that was certainly going to appeal to a wider audience – sales figures, of course, went on to prove that. As with the 2+2 version of the 260Z, the new car's longitudinal stability was an improvement on the shorter two-seater and the rear seat occupants were less likely to feel encapsulated than before, as headroom was marginally improved and the seats themselves were just a little more comfortable. Most pleasing of all the 280ZX 2+2's features was undoubtedly the fact that it turned in a pretty similar performance to the two-seater.

In Britain, former racing driver John Miles in his regular column in *Autocar* was quite critical of the ZX. In opening his review he said:

> Anybody looking back over Z-car history might wonder why what started out as a light, fast, compact and tolerably economical sports car the 240Z – has through the 260 and now 280 series become progressively slower, not at all economical, larger – for no apparent increase in inside space – also heavier and relatively expensive.

To be fair, his review was not all in that vein; he did have some good things to say. He clearly had liked the 240Z and had not welcomed its development to comply with American regulations. But he recognized that it was essential, in the light of the greater sales potential there, so went on to find the good points of the 280ZX. He commented that in road grip and manners the new Datsun beat the Ford Capri 3000S hands down (no surprise there) and was pretty close, if not equal, to the Porsche 924. He finished by saying that, well appointed and comfortable though the 280ZX was, the T-bar open-topped 280ZX Turbo, which had its debut at the Frankfurt Motor Show, confirmed that Nissan knew what was needed to turn it back into a sports car: 50bhp!

Comparative road tests were now much more fashionable in magazines both in Britain and in the United States, so more editors were looking for road tests that not only told how a car performed and what its shortfalls were, but how those characteristics stacked up against the competition. So in June 1979 *Motor Trend* published a comparison of the 280ZX 2+2 and the Toyota Celica Supra. Despite criticisms offered for the 'stretch' of the 280ZX to create the 2+2 that resulted in an elongated line and disruption of the natural roof curvature, the Datsun was distinctly the more attractive to look at, for the Toyota was a combination of mismatched angles and curves, though some people must have loved it, for it also sold in quite large numbers.

Out on the road, the Toyota was found to be smooth in just about all aspects. The controls were all immediately to hand and operated 'with a silky feel'. The engine was an instant starter and idled smoothly, picking up quickly under acceleration, providing good power and torque, propelling the car to 60mph in 11.5 seconds on an engine of 2,563cc and 110bhp. The all-round disc

HOW THE DATSUN 280ZX 2+2 AUTOMATIC PERFORMS

Figures taken at 6,800 miles by our own staff at the Motor Industry Research Association proving ground at Nuneaton.

All Autocar test results are subject to world copyright and may not be reproduced in whole or part without the Editor's written permission.

TEST CONDITIONS:
Wind 8-20 mph
Temperature 18 deg C (64 deg F)
Barometer 29.2 in. Hg (990 mbar)
Humidity 54 per cent
Surface dry asphalt and concrete
Test distance 1,063 miles

MAXIMUM SPEEDS

Gear	mph	kph	rpm
Top (mean)	111	179	5,500
(best)	114	183	5,650
2nd	88	141	6,400
1st	52	84	6,400

ACCELERATION

FROM REST

True mph	Time (sec)	Speedo mph
30	4.1	32
40	5.8	43
50	8.2	53
60	11.3	64
70	15.2	74
80	20.9	84
90	28.5	95
100	39.9	107
110	—	119

Standing ¼-mile: 18.3 sec, 75½ mph
Standing km: 33.7 sec, 95½ mph

IN EACH GEAR

mph	Top	2nd	1st
0-20	—	3.5	2.5
10-30	—	4.4	3.2
20-40	—	5.5	3.4
30-50	—	5.8	4.1
40-60	—	5.8	—
50-70	—	6.6	—
60-80	—	9.2	—
70-90	12.8	—	—
80-100	19.7	—	—

FUEL CONSUMPTION

Overall mpg: 18.4 (15.4 litres/100km)
Constant speed: figures not available; car's fuel injection system incompatible with Autocar petrol flow meter
Autocar formula Hard 16.6 mpg
Driving Average 20.2 mpg
and conditions Gentle 23.9 mpg

Grade of fuel Regular, 2-star (91 RM)
Fuel tank 17.6 Imp. galls (80 litres)
Mileage recorder reads 2.3 per cent long

Official fuel consumption figures
(ECE laboratory test procedures, not necessarily related to Autocar figures)
Urban cycle 19.3 mpg
Steady 56 mph 29.7 mpg
Steady 75 mph 24.1 mpg

OIL CONSUMPTION
(SAE 20W/50) 1,100 miles/pint

BRAKING

Fade (from 76 mph in neutral)
Pedal load for 0.5g stops in lb

	start/end		start/end
1	24-18	6	30-60
2	20-16	7	35-50
3	20-18	8	35-70
4	24-20	9	35-55
5	28-40	10	30-55

Response (from 30 mph in neutral)

Load	g	Distance
10lb	0.20	150ft
20lb	0.40	75ft
30lb	0.62	49ft
40lb	0.80	38ft
50lb	0.92	32.8ft

Handbrake 84ft
Max gradient: 1 in 3

WEIGHT
Kerb, 25.7 cwt, 2,881 lb/1,307 kg
(Distribution (F/R, 51.0/49.0)
Test, 29.3 cwt, 3,281 lb/1,488 kg
Max. payload 880 lb/400 kg

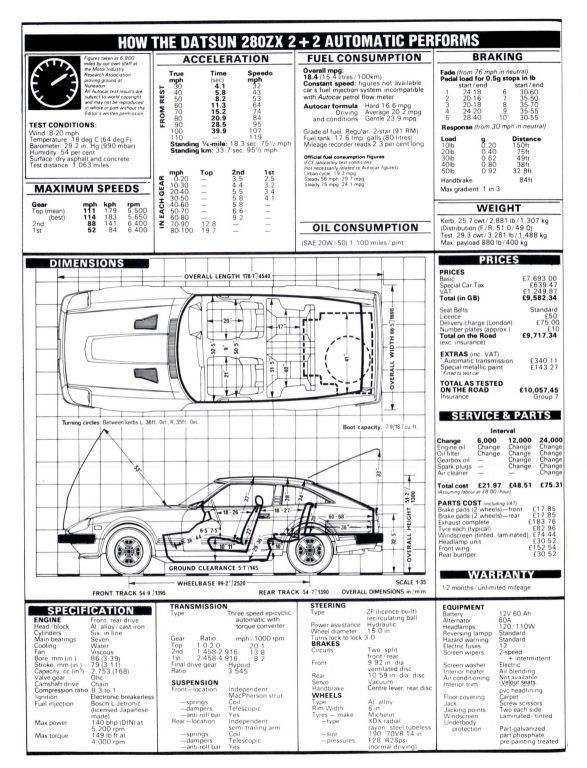

DIMENSIONS
OVERALL LENGTH 178·7″/4540
OVERALL WIDTH 66·5″/1690
OVERALL HEIGHT 51·2″/1300
WHEELBASE 99·2″/2520
FRONT TRACK 54·9″/1395
REAR TRACK 54·7″/1390
GROUND CLEARANCE 5·7″/145
Turning circles: Between kerbs L, 36ft. 0in, R, 35ft. 0in.
Boot capacity: 7·9/18·7 cu.ft.
SCALE 1:35
OVERALL DIMENSIONS in/mm

PRICES

Basic	£7,693.00
Special Car Tax	£639.47
VAT	£1,249.87
Total (in GB)	**£9,582.34**
Seat Belts	Standard
Licence	£50
Delivery charge (London)	£75.00
Number plates (approx.)	£10
Total on the Road	**£9,717.34**
(exc. insurance)	

EXTRAS (inc. VAT)
* Automatic transmission £340.11
Special metallic paint £143.27
Fitted to test car

TOTAL AS TESTED ON THE ROAD £10,057.45
Insurance Group 7

SERVICE & PARTS

	Interval		
Change	6,000	12,000	24,000
Engine oil	Change	Change	Change
Oil filter	Change	Change	Change
Gearbox oil	—	Change	Change
Spark plugs	—	Change	Change
Air cleaner	—	—	Change
Total cost	**£21.97**	**£48.51**	**£75.31**

(Assuming labour at £8.00/hour)

PARTS COST (including VAT)
Brake pads (2 wheels) —front £17.85
Brake pads (2 wheels) —rear £17.85
Exhaust complete £183.76
Tyre each (typical) £82.96
Windscreen (tinted, laminated) £74.44
Headlamp unit £30.52
Front wing £152.54
Rear bumper £30.52

WARRANTY
12 months/unlimited mileage

SPECIFICATION

ENGINE
Head/block Front, rear drive
Head/block Al. alloy/cast iron
Cylinders Six, in line
Main bearings Seven
Cooling Water
Fan Viscous
Bore, mm (in.) 86 (3.39)
Stroke, mm (in.) 79 (3.11)
Capacity, cc (in³) 2,753 (168)
Valve gear Ohc
Camshaft drive Chain
Compression ratio 8.3-to-1
Ignition Electronic breakerless
Fuel injection Bosch L-Jetronic (licensed Japanese-made)
Max power 140 bhp (DIN) at 5,200 rpm
Max torque 149 lb ft at 4,000 rpm

TRANSMISSION
Type Three-speed epicyclic automatic with torque converter

Gear	Ratio	mph/1000 rpm
Top	1.0-2.0	20.1
2nd	1.458-2.916	13.8
1st	2.458-4.916	8.2

Final drive gear Hypoid
Ratio 3.545

SUSPENSION
Front—location Independent, MacPherson strut
 —springs Coil
 —dampers Telescopic
 —anti-roll bar Yes
Rear—location Independent, semi-trailing arm
 —springs Coil
 —dampers Telescopic
 —anti-roll bar Yes

STEERING
Type ZF (licence-built) recirculating ball
Power assistance Hydraulic
Wheel diameter 15.0 in.
Turns lock to lock 3.0

BRAKES
Circuits Two, split front/rear
Front 9.92 in. dia ventilated disc
Rear 10.59 in. dia disc
Servo Vacuum
Handbrake Centre lever, rear disc

WHEELS
Type Al. alloy
Rim Width 6 in
Tyres — make Michelin
 —type XDX radial rayon steel tubeless
 —size 190 70VR 14 in
 —pressures F28 R28psi (normal driving)

EQUIPMENT
Battery 12V 60 Ah
Alternator 60A
Headlamps 120/110W
Reversing lamp Standard
Hazard warning Standard
Electric fuses 12
Screen wipers 2-speed + intermittent
Screen washer Electric
Interior heater Air blending
Air conditioning Not available
Interior trim Velour seats, pvc headlining
Floor covering Carpet
Jack Screw scissors
Jacking points Two each side
Windscreen Laminated/tinted
Underbody protection Part-galvanized, part phosphate pre-painting treated

Autocar's test data for the 280ZX 2+2.

ROAD TEST DATA
Datsun 280ZX 2+2

SPECIFICATIONS

GENERAL
Vehicle type	Front-engine, rear-drive, 2 + 2 hatchback sport coupe
Base price	$11,599
Options on test car	Grand Luxury package
Price as tested	$13,203

ENGINE
Type	Inline six, overhead cam, water cooled, cast iron block
Bore & stroke	3.39 x 3.11 in.
Displacement	168.0 cu. in.
Compression ratio	8.3:1
Fuel system	Electronic fuel injection
Recommended octane number	91 RON
Emission control	Calif.
Valve gear	Overhead cam
Horsepower (SAE net)	135 at 5200 rpm
Torque (SAE net)	144 lb.-ft. at 4000 rpm
Power to weight ratio	21.6 lb./hp

DRIVETRAIN
Transmission	Manual 5-speed
Final drive ratio	3.7:1

DIMENSIONS
Wheelbase	99.2 in.
Track, F/R	54.5/54.9 in.
Length	181.9 in.
Width	66.5 in.
Height	51.4 in.
Ground clearance	5.9 in.
Max. load length w/rear seat(s) folded down	54.5 in.
Curb weight	2915
Weight distribution, F/R	TK

CAPACITIES
Fuel capacity	21 gals.
Crankcase	4.8 qts.
Cooling system	11.1 qts.
Cargo volume	18.2 cu. ft.

SUSPENSION
Front	MacPherson strut, tension rods, coil springs, stabilizer bar, double-acting shock absorber
Rear	Independent, MacPherson strut, semi-trailing arms, coil springs, stabilizer bar, double-acting shock absorbers

STEERING
Type	Power-assisted, variable-ratio rack and pinion
Turns lock-to-lock	2.7
Turning circle, curb-to-curb	34.8 ft.

BRAKES
Front	Power-assisted ventilated 9.9-in. discs
Rear	Power-assisted 10.6-in. discs

WHEELS AND TIRES
Wheel size	6 x 14 in.
Wheel type	Aluminum alloy
Tire make and size	Bridgestone 195/70 HR14
Tire type	Steel radial

TEST RESULTS

Recommended pressure, F/R	28/28 psi

ACCELERATION
0-30 mph	3.3 secs.
0-40 mph	5.5 secs.
0-50 mph	7.7 secs.
0-60 mph	11.3 secs.
0-70 mph	15.0 secs.
0-80 mph	19.5 secs.
Standing quarter mile	18.0/77 mph

BRAKING
30-0 mph	34 ft.
60-0 mph	160 ft.

FUEL CONSUMPTION
EPA city	18.0 mpg
MT 73-mile test loop	19.4 mpg

SPEEDOMETER
Indicated	30	40	50	60
Actual mph	29	38	47	56

Motor Trend *produced this fascinating comparison of the 280ZX and the Toyota (opposite).*

ROAD TEST DATA
Toyota Celica Supra

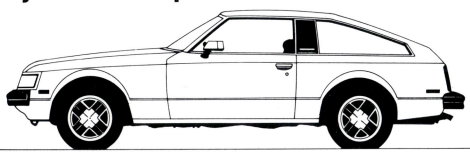

SPECIFICATIONS

GENERAL
Vehicle type	Front-engine, rear-drive, 4-pass. hatchback sport coupe
Base price	$9578
Options on test car	Black paint, rear window wiper, cruise control
Price as tested	$9850

ENGINE
Type	Inline six, overhead cam, water-cooled, cast iron block
Bore & stroke	3.15 x 3.35 in.
Displacement	156.4 cu. in.
Compression ratio	8.5:1
Fuel system	Electronic fuel injection
Recommended octane number	91 RON
Emission control	Calif.
Valve gear	Overhead cam
Horsepower (SAE net)	110 at 4800 rpm
Torque (SAE net)	136 lb.-ft. at 2400 rpm
Power to weight ratio	25.6 lb./hp

DRIVETRAIN
Transmission	Manual 5-speed
Final drive ratio	3.9:1

DIMENSIONS
Wheelbase	103.5 in.
Track, F/R	53.7/53.7 in.
Length	181.7 in.
Width	65.0 in.
Height	51.8 in.
Ground clearance	5.7 in.
Max. load length w/rear seat(s) folded down	65.8 in.
Curb weight	2855 lb.
Weight distribution, F/R	53/47%

CAPACITIES
Fuel capacity	16.1 gals.
Crankcase	5.5 qts.
Cooling system	11.6 qts.
Cargo capacity	24.3 cu. ft.

SUSPENSION
Front	MacPherson strut, coil springs, stabilizer bar, double-acting shock absorbers
Rear	Live axle, 4-bar link, stabilizer bar, double-acting shock absorbers

STEERING
Type	Power-assisted recirculating ball
Turns lock-to-lock	3.4
Turning circle, curb-to-curb	34.8 ft.

BRAKES
Front	Power-assisted 10.0-in. discs
Rear	Power-assisted 8.6-in. discs

WHEELS AND TIRES
Wheel size	5.5 x 14 in.
Wheel type	Stamped steel
Tire make and size	Bridgestone 195/70 HR14
Tire type	Steel radial
Recommended pressure, F/R	24/24 psi

TEST RESULTS

ACCELERATION
0-30 mph	3.6 secs.
0-40 mph	5.7 secs.
0-50 mph	8.3 secs.
0-60 mph	11.5 secs.
0-70 mph	15.7 secs.
0-80 mph	21.5 secs.
Standing quarter mile	18.4 secs/74.6 mph

BRAKING
30-0 mph	35 ft.
60-0 mph	168 ft.

FUEL CONSUMPTION
EPA city	19 mpg
MT 73-mile test loop	20.7 mpg

SPEEDOMETER
Indicated	30	40	50	60
Actual mph	28	38	48	58

Total Production of Datsun/Nissan Sports Cars 1958 to 1993

Year	Model Name	Model Number	Quantity Built
1958	Datsun Sports	S211	3
1959	Datsun Sports	S211	14
	Fairlady Sports	SP212	288
1960	Datsun Sports	S211	3
	Fairlady Sports	SP213	46
1961	Fairlady Sports	SP213	199
1962	Fairlady Sports	SP213	52
	Fairlady/1500 Sports	SP310	187
1963	Fairlady/1500 Sports	SP310	1,977
1964	Fairlady/1500 Sports	SP310	4,550
	Fairlady/1600 Sports	SP311	6
1965	Fairlady/1500 Sports	SP310	192
	Fairlady/1600 Sports	SP311	4,774
1966	Fairlady/1600 Sports	SP311	6,103
	Fairlady/2000 Sports	SR311	2
1967	Fairlady/1600 Sports	SP311	5,452
	Fairlady/2000 Sports	SR311	2,210
1968	Fairlady/1600 Sports	SP311	6,136
	Fairlady/2000 Sports	SR311	7,554
1969	Fairlady/1600 Sports	SP311	3,759
	Fairlady/2000 Sports	SR311	5,109
	Fairlady Z/240Z	HS30	1,162
1970	Fairlady 1600 Sports	SP311	1,154
	Fairlady 2000 Sports	SR311	131
	Fairlady Z/240Z	HS30	21,837
1971	200Z/240Z	HS30	44,998
1972	200Z/240Z	HS30	65,956
1973	200Z/240Z	HS30	58,596

Year	Model Name	Model Number	Quantity Built
1974	200Z/240Z/260Z	S 30	62,961
1975	280Z/Fairlady Z	S 30	72,503
1976	260Z/280Z/2 + 2/Fairlady Z	S 30	72,565
1977	260Z/280Z/2 + 2/Fairlady Z	S 30	84,156
1978	260Z/280Z/2 + 2/Fairlady Z	S 30	46,867
	280ZX/2 + 2/Fairlady Z	S 130	44,353
1979	280ZX/2 + 2/Fairlady Z	S 130	105,045
1980	280ZX/2 + 2/Fairlady Z	S 130	70,435
1981	280ZX/2 + 2/Fairlady Z	S 130	84,668
1982	280ZX/2 + 2/Fairlady Z	S 130	74,030
	300ZX/Fairlady Z	Z-31	5
1983	280ZX/2 + 2/Fairlady Z	S 130	36,097
	300ZX/2 + 2/Fairlady Z	Z-31	36,555
1984	300ZX/2 + 2/Fairlady Z	Z-31	96,346
1985	300ZX/2 + 2/Fairlady Z	Z-31	78,765
1986	300ZX/2 + 2/Fairlady Z	Z-31	66,265
1987	300ZX/2 + 2/Fairlady Z	Z-31	37,888
1988	300ZX/2 + 2/Fairlady Z	Z-31	12,775
1989	300ZX/2 + 2/Fairlady Z	Z-31	14,296
	300ZX/2 + 2/Fairlady Z	Z-32	36,163
1990	300ZX/2 + 2/Fairlady Z	Z-32	42,563
1991	300ZX/2 + 2/Fairlady Z	Z-32	26,313
1992	300ZX/2 + 2/Fairlady Z	Z-32	19,075
1993	300ZX/2 + 2/Fairlady Z	Z-32	8,750
		Total	**1,471,889**

brakes brought the car to a stop in a straight line from 30mph (48kph) in 35ft (11m). The fractionally longer wheelbase of this car seems to have endowed it with a better ride under braking and acceleration than the Datsun, though much of that also was down to the design and ratings of the suspension, of course. Overall, *Motor Trend* drew the conclusion that in many ways the Toyota was a better car for ride and general comfort, though it needed a bit of attention to the rear suspension to keep the axle on the road on a rough surface and it needed a bit more power. On the other hand, there was now a strong and loyal band of Z enthusiasts who would see the Datsun through its problems and buy it for emotional reasons.

'Orient Express' is how *Car and Driver* described the 280ZX, which delivered the fastest ride of any Japanese car yet. *Motor Trend* pushed the Turbo past 60mph in under 8 seconds and managed to bring it to a standstill again from that speed in a distance of 150ft (46m). Not bad – now Nissan were producing a sports car again. Despite the extra weight, here was a car that was a true handler in the spirit of the original 240Z. Its rivals were such cars as the Porsche 924 Turbo and the Alfa Romeo Alfetta GTV6. This was the ultimate development of the in-line six cylinder engined Z-car.

NEW ENGINE, NEW SHAPE – THE 300ZX

At the end of 1983, the name 'Nissan' began to appear on all cars made by the company and 'Datsun' disappeared in the process of corporate consolidation. With the new name came a new car, the 300ZX, taking another step up the ladder of luxury on the way. This new V-6 engined 3 litre machine was about as far removed from the lean, almost Spartan 240Z two-seater of 14 years before as it could

be. Retaining a shape similarity to the 280ZX for the sake of preserving the family line was a conscious decision and the press liked that, but they liked even more the new V-6 power unit and what it brought to the new car.

Road and Track saw this new Nissan as a genuine rival for the new Corvette from Chevrolet, based not on engine size, but on pure appointments and performance. The T roof, with its lockable panels, was now seen as an attractive feature, whereas a number of journalists had not been sure with the previous model. *Road and Track* went on to compare the 300ZX with the Corvette and the Porsche 928. When comparing the driving feel between those three cars, it went so far as to comment that the 300ZX was perhaps a little more lithe than the other two, which you might expect as it did not carry as much bulk, though the solid feel of the car was part of that comment.

The cockpit appointment was thought to be superb, as was the general road performance of the 300ZX. Its road manners were generally excellent, though the tester found it a bit tail happy and inclined to roll a little. In a straight line, it could reach 60mph in 7.4 seconds and stop again from that speed in 141ft (43m). Maximum speed was recorded as 133mph (214kph) and fuel consumption was a remarkable 20 to 28mpg (10 to 14 litres per 100km), depending where and how you drove it. The tester's conclusion was that the new 300ZX was not the same revolution in the market place as was the 240Z, but that it would still make a significant impact on its own market.

The real revolution came with the new 300ZX – the next (and in Britain the final) version of the Z, just 20 years after the 240Z took the sports car world by storm. This was a new, beautifully proportioned, aerodynamically clean vehicle that was again far removed from the first-of-the-line 240Z. Now we had a car that compared well and

Road Tests and Press Reviews

The 1983 Corvette (above), rival in America for the 300ZX, and the Z-31 Series new 300ZX Targa (below).

directly with the Porsche 944, as well as several other top-of-the-market models. The North American market was to receive the two-seater and the 2+2, the two-seater being the only version available with the Turbo option there, while in Europe the car was to be offered in 2+2 Turbo form only. The press clamoured to see and drive this new car and the buying public just stood in line and waited.

Car and Driver was one of the first American journals to put the new 300ZX through its paces and they found the 'sport' setting for the shock absorbers to be too hard, compared with the smoother ride of the Porsche and the Corvette. The steering was found to be precise and sensitive without passing all the vibrations of a less than smooth surface back to the driver. For *Autocar* in Britain, Ray Hutton found that same steering feature and praised it, but *Car* thought it grounds for criticism. The reviewer found the acute sensitivity of the Porsche 944 to be far preferable, as he could feel every pebble on the road – he thought the Nissan's tendency to absorb it all, even though its steering was every bit as precise as the Porsche, failed to show the driver just how cleverly it was doing its job!

Within a couple of years of the announcement of the new 300ZX a small company called Motor Sports International, based in

Road Tests and Press Reviews

The 300ZX (Z-32) out on the road.

Waco, Texas, put a developed version of the car into the hands of the American magazine *Automobile* for test. It was called the Nissan 300ZX Turbo SR-71 and the twin turbochargers were 'tweaked' in conjunction with the electronics that controlled ignition and fuel supply to produce an astounding 428bhp! What was more, that power output was achieved in compliance with even California's exhaust emission standards. The inevitable body 'add-ons' came from Kaminari, the tyres were changed for Goodyear 245/45-17s on the front and 275/40-17s at the rear. The reviewer described the resultant performance and handling as 'stunning' but gave no actual figures, so you are left to imagine how it would have felt, though his parting shot tells you that the Nissan 300ZX will never feel the same again. It had a price tag of around $68,000.

Finally, the one car that everyone said Nissan would never build came in 1992 – the 300ZX Convertible. The press loved it, not least because it was a very clever application of the obvious, in that the process of making the T-roof coupé a convertible came by simply removing the T bar and the huge clamshell tailgate. In performance and handling it behaved just like the naturally aspirated coupé; though the convertible version would not sell in huge numbers, the value of offering the option was seen as advantageous to Nissan's marketing strategy. Little 'tweaks' like cornering lamps were now included in the specification, while such safety features as an anti-lock braking system and a viscose limited slip differential continued into this new car, too. Road noise was reckoned not to be a problem at the Interstate 65mph (105kph) limit and most reporters thought this to be a natural convertible, as distinct from some of the rather contrived devices coming on to the road with the growth in popularity of open cars in general.

It is sad that 1994 should have seen the demise of the Z in Britain. In a quarter of a century, Datsun and Nissan cars had established an enviable reputation for innovation, quality, performance and value that was becoming very hard to beat.

11 Proving the Z in Racing and Rallying

One of the great assets of the Datsun and Nissan Z models has been their durability and sheer mechanical strength. That, combined with the manufacturer's consistent determination to show the world it made quality cars, ensured that whenever a Datsun or Nissan went to do battle on the roads and tracks of rallying and racing, it would be noticed. From the very earliest successes of the 240Z in rallies in Europe and Africa, then races in North America, to the latest GTU competitions in the USA with the 300ZX, Nissan has shown that it can make a vehicle to do the job of winning.

DATSUN'S EARLY RALLYING AND RACING

In the United States, the Datsun 1500, 1600 and 2000 sports cars acquitted themselves well in club racing during the 1960s and established a reputation for solid cars that could hold their own. Some of these cars also took part in rallies, but the most significant breakthrough in setting any kind of sporting reputation for the company was a plain Datsun 210 saloon, which ploughed its way to a class win in the 1958 Australian Mobilgas Trial, a rally which ran round the

Out on the track, the 1600 busily made its mark and began to beat the European competition in production sports car racing.

world's largest island and covered some of the world's roughest terrain. That little car did more for the reputation of Datsun products in Australia than any amount of ordinary road reliability or sales campaigning could ever do. The Mobil run experience also provided Nissan with the foundation for a sports car programme that would give it credit for the world's largest volume mass-produced sports car inside a dozen years.

Setting about establishing a reputation for winning sports cars demanded a lot and Datsun USA began by finding the best skills in North America to launch a race programme that would not reflect too badly on the manufacturer if it failed, but would bring tremendous charisma if it succeeded. The technique was to find that skill to work in a relatively low key situation, so as not to attract too much publicity too soon, and to put all the money it needed, within reason, at the disposal of the project. The occasional early win would secure a little respect, without causing a major focus of attention on Datsun. Nobody wanted to alert the competition before that winning streak was pretty firmly established. So the names John Morton and Bob Sharp were recruited to run under the Datsun flag. This would guarantee the right publicity in a winning situation, but not attract too much attention to Datsun cars if they failed.

The philosophy adopted in this early attack on the United States motor sporting scene was not exactly clandestine, but was part of the process of a very clever programme to secure a large chunk of America's sports car sales business, displacing many British manufacturers in the process. Honda had followed a similar programme in the 1950s in the motor cycle world. There, Soichiro Honda had set himself the target of winning 125cc and 250cc European motor cycle races in 1959. Within a couple of years, Honda had displaced all comers to become supreme in those two classes and had become an international name in the world of two wheels, producing some 20 per cent of the whole world's motor cycles by 1962. The continuing decline of the British motor cycle industry brought the name Honda to the pinnacle by the end of the decade. Nissan was aiming for a similar goal, at least in North America, with its sports cars.

It is fair to say that Datsun sports cars did not often make the finish line in their early forays into American sports car racing. Perhaps it was not too surprising, in a way, as they were taking on the cream of European engineering in the form of Alfa Romeos, Austin Healeys, Lotus, MGs and Triumphs – all out there with nothing less than winning in mind. This racing experience also helped to develop the character and performance potential of the production cars, taking the 1500 Sports from 96bhp through to the 2000's 150bhp with the twin Solex carburettor option. Using the lessons of that early American experience, Nissan slowly got the engineering aspect right and began winning races. Then they got into the habit of winning races and with the development work done by their American race team experts, they were ready for the next step – the entry of the 240Z into the competition scene.

VICTORY IN SIGHT

The next link in the chain to true international success for Nissan was the Datsun 1600SSS, a car that even today strikes dread into the heart of the hardened rally driver. For it was the car which brought the name Datsun to the forefront of the international rally scene and paved the way for the 240Z. The East African Safari Rally was the place where they finally broke through. Britain's premier sporting journal *Motor Sport* made the point:

It was never really doubted that one day a Datsun would win the East African Safari Rally. For several years, the Japanese team has put all its efforts into this tough event, only tackling the odd European rally when available time happened to coincide with the inclination.

That statement really put the Japanese car maker's objectives into perspective. Edgar Herrmann and Hans Schuller, a very respected and experienced rally team, had beaten the establishment and finally won, trailed by Jovinder and Ranyard, two local drivers, into second place. Datsun had done it – they were now true players on the international scene.

The Safari Rally began as a special event to mark the 1953 Coronation. It was organized by a group of British expatriates and covered 2,000 miles of some of the roughest tracks imaginable, for there were few metalled roads anywhere in East Africa in those days. Over the years the event became more sophisticated and better organized to reach the standard of an international rally, vying for status with such other major items on the calendar as the Acropolis, the Thousand Lakes or even the Alpine. One aspect of the Safari that set it apart from the rest was its unpredictability, for it was just not possible to set a fixed route, as weather changes could flood or eliminate a road, so alternatives had to be selected for last minute changes.

The 1970 Safari Rally was a tough one, covering almost 4,000 miles (6,400km), with injuries and retirements taking more than their usual toll of the entrants. For example, a Japanese crew were entered to drive a 1600SSS Datsun and one of the pair was killed in practice. Then another Datsun crew, Simonian and Neylan, who did not have a complete set of route notes, crashed through the parapet of a bridge and landed upturned in the dry river bed below. They retrieved the car, carried on their drive, but retired 1,000 miles (1,600km) later with a loss of oil when the filter casing worked loose. Despite all this, seven of the nine entries finished, three of them in the top four, securing for Datsun the Team Prize, beating the products of Alfa Romeo, Peugeot, SAAB, Toyota, Triumph and Volvo on the way.

The last major international European rally of 1970 was the RAC in Britain. There was a lot of pressure on the Datsuns here, as the BMC Minis and Ford Escorts were on home ground, quite apart from stiff opposition being mounted by Lancia – who won that year with Harry Kallstrom and Gunnar Haggbom in a Fulvia 1600HF. Opel took the Team Prize, finishing 2nd, 3rd and 4th and taking the first three places in Class Four. But in 7th place came a real surprise. It was the diminutive Finnish driver, Rauno Aaltonen, partnered by Paul Easter, driving a Datsun 240Z, the first European outing for the new Japanese sports car that had received such acclaim in the United States after its release into that market at the beginning of the year. Already, by the time the RAC Rally took place, the 240Z had been voted 'Imported Sports Car of the Year' in the United States.

The scent of victory was in the nostrils of Nissan's competition department now, as the Sports Car Club of America's Class C championship came within their grasp for sports car racing on the tracks of America. So now, they had won one of the world's most gruelling rallies and a major racing championship against some of the toughest opposition available. Then, in 1971, they did it again on the rally front by coming home in 5th and 10th in that year's Monte Carlo Rally, going on to win the East African Safari for a second year, with the same winning crew of Herrmann and Schuller. Another local crew, Mehta and Doughty, finished the Safari second, with a third 240Z, crewed by Aaltonen and Easter, coming home 7th and so securing the Team Prize again for Datsun.

Proving the Z in Racing and Rallying

Herrmann and Schuller won the 1971 East African Safari with a 240Z, whilst Rauno Aaltonen and Paul Easter crewed one of the 1972 Monte Carlo Rally cars to finish third.

Proving the Z in Racing and Rallying

BUILDING ON A SOLID FOUNDATION

More races, more rallies – the 1972 season continued to bring victories and major placings to Nissan's cars and the United States market was now truly in love with the car. Sports car enthusiasts in Europe were becoming more impressed with the 240Z, too, as in January it entered the first major international rally of the year – the 41st Monte Carlo. A spectacular 3rd place overall was the result, then at the end of March, it was time for another East African Safari, though the victory they had hoped for eluded them this time, with only 5th, 6th and 10th for the 240Zs. A near victory in Australia's Southern Cross Rally saw a Datsun win its class, but take only second place overall. The TAP Rally in Portugal brought a 4th place, while back to Britain for the RAC event saw a 240Z placed 11th, with another Class win. Across the Atlantic the 240Z did well in rallying too, as the SCCA Pro Rally Manufacturer's Championship went to Datsuns, as well as a third SCCA National racing Championship for Production Sports Cars in Class C.

So 1972 was another good year in rallying for Datsun, with wins in the Portuguese Montana Rally, the Welsh Rally, the Kenya 2000 and many minor rallies. Then, there were other significant placings, which included a 19th place in the RAC of that year. Racing brought the Japanese GT Grand Prix, the Singapore Grand Prix GT Races, the Selangor Grand Prix Manufacturer's Championship, another SCCA Class C Sports Car Championship with Bob Sharp Racing leading the field as Sam Posey and Paul Newman piloted 240Zs to victory after victory. The Japanese Grand Prix for sports and GT cars saw another Datsun victory, probably the most significant, being on home ground and demonstrating to a home crowd how good and reliable a car the 240Z was.

Now, the task for Datsun USA was to keep the momentum up, as the 260Z replaced the 240Z in the market place and as Z models

Shekhar Mehta and Lofty Drew brought another East African Safari victory to Nissan with the Datsun 240Z in 1973.

faced stiffening competition on rally roads and race tracks. Another Monte Carlo brought a 9th place in 1973, while victory was not elusive in that year's Safari Rally in East Africa. In August, the Tanzania 1000 Mile Rally also went to a Datsun, while a modest 14th was the best the Z could do in the RAC. Again, the race tracks were popular places for the Z to be seen, as pace cars for the organizers and as participants. Every year between 1973 and 1977, the SCCA National Championships for Production Cars in Class C fell to Datsuns. The IMSA GTU Drivers' Championships also went to Datsun drivers between 1974 and 1977, while the Manufacturers' Championship was secured by Datsun Z models in 1975, 1976 and 1977. Pro Rally Championships fell to Datsuns and finally, the North American Rally Cup Manufacturers' Championship went that way too.

In November 1975, *Road and Track* published a very unusual test report of 'a Pair of Z Cars', the pair being a production 280Z and one of Bob Sharp's race prepared 240Zs. The test was conducted at Lime Rock Park in Connecticut. Setting qualifying and lap records in the previous day's IMSA 'Grand Touring Under 2.5 Litre' class, the 240Z had won handsomely and was now to be pitted against the fastest selling sports car in America. Bob Sharp knew more than a little about Z cars, having won thirty national races with one and two SCCA Championships at the time the article was published. The two cars were as different as chalk from cheese, but the purpose of the track test was to show the extremes to which a Z could be taken.

Earlier in that year, 1975, *Road and Track* had chosen the 280Z as best car in its class in their 'Ten Best Cars for a Changing World' selection. It had also been voted 'Imported Sports Car of the Year' elsewhere and was doing quite a lot to restore the original sporting charisma of the 240Z of six years earlier. Bob Sharp, of course, owned Bob Sharp Datsun and he provided the 280Z for this test. It was a perfectly factory-standard car that was new, but run in ready for the comparison test. Unlike all the other Datsuns Bob Sharp had driven on the track, this one did not even have a heavier anti roll bar or modified shock absorbers. This would be the first time he had driven an absolutely catalogue standard car round Lime Rock.

On the other hand, the racing car was about as race prepared as it could be, having come fresh from its previous day's race win on that track. Bob Sharp Racing's 240Z was about as far removed from catalogue specification as it was possible to make it. The engine now had a displacement of 2,450cc, bored out to bring it closer to the 2.5-litre limit of its class. Fuelled by three Mikuni-Solex carburettors (not fuel injection), the compression ratio was increased to 11.8:1, almost diesel levels. Power output of this much-modified 'mill' was more than 100bhp over the standard rating of the catalogue engine, giving 258bhp at 8,250rpm. Even at this, that power, transmitting through a five-speed gearbox, only managed to propel the car to 60mph in 7 seconds, though the car still weighed 2,200lb (998kg). Much of that was from the interior roll cage, a device that seemed to make it impossible to get into the car at all!

Given that the anti roll bars were heavier duty and that the adjustable shock absorbers were not what were fitted to a standard 240Z, the suspension was quoted as being 'basically stock', which meant that all the bits looked like those fitted to a production car, but were not quite the same! Various adjustments were made to improve roll stability, while wheel and tyre equipment increased the overall width of the car by over a foot. On the front were 15in diameter by 10in wide rims, carrying 24.5 × 10/15

Proving the Z in Racing and Rallying

Bob Sharp Racing fielded two much-modified 'Z'-cars in SCCA racing in the United States for Sam Posey and Paul Newman.

Proving the Z in Racing and Rallying

SPECIFICATIONS COMPARISON
Production & Racing Datsun Z Cars

	Production	Racing
Price	$6284	$25,000
General:		
Weight, lb	2875 (curb)	2200 (racing)
Weight distribution (with driver), front/rear, %	50/50	52/48
Track, front/rear, in.	53.3/53.0	58.0/63.5
Length, in.	173.2	171.0
Width, in.	64.2	82.0
Height, in.	51.0	49.0
Ground clearance, in.	5.7	3.6
Usable trunk space, cu ft	8.5	nil
Fuel capacity, U.S. gal.	17.2	22.0
Engine:		
Bore x stroke, mm	86.1 x 79.0	84.0 x 73.7
Displacement, cc/cu in.	2754/168	2450/149
Compression ratio	8.3:1	11.8:1
Bhp @ rpm, net	149 @ 5600	258 @ 8250
Torque @ rpm, lb-ft	163 @ 4400	184 @ 6500
Carburetion/fuel injection	L-Jetronic injection	three Mikuni-Solex (2V)
Fuel requirement	regular, 91-oct	premium, 102-oct
Drive train:		
Transmission	4-sp manual	5-sp manual
Gear ratios:		
5th	—	1.00
4th	1.00	1.14
3rd	1.31	1.29
2nd	2.08	1.60
1st	3.32	2.35
Final drive ratio	3.55:1	4.11:1
Chassis:		
Brake system	10.7-in. discs front, 9.0 x 1.6-in. drums rear; vacuum assisted	10.7-in. discs front and rear
Swept area, sq in.	310	394
Wheels	steel disc, 14 x 5J	Minilite 10 x 15 front, BBS 15 x 15 rear
Tires	Bridgestone, Wide 70 195/70-14	Goodyear Blue Streak 24.5 x 10-15 front, 25.0 x 13-15 rear
Front suspension	MacPherson struts, lower lateral arms, compliance struts, coil springs, tube shocks, anti-roll bar	MacPherson struts, lower lateral arms, compliance struts, coil springs, adj tube shocks, adj anti-roll bar
Rear suspension	Chapman struts, A-arms, coil springs, tube shocks, anti-roll bar	Chapman struts, A-arms, coil springs, adj tube shocks, adj anti-roll bar
Instrumentation		
Instruments	160-mph speedo, 8000-rpm tach, 99,999 odo, 999.9 trip odo, oil press., coolant temp, ammeter, fuel level, clock	10,000-rpm tach, oil press., oil temp, coolant temp, ammeter, differential temp, transmission temp
Warning lights	brake system, rear-window heat, seatbelts, egr, hazard, high beam, directionals	oil pressure
Accommodation:		
Seating capacity, persons	2	1
Seat width, in.	2 x 21.0	14.0
Head room, in.	36.5	38.0
Seat back adjustment, deg	40	0
Calculated data:		
Lb/bhp (test weight)	21.2	9.8
Mph/1000 rpm	20.3 (4th gear)	17.6 (5th gear)
Engine revs/mi (60 mph)	2960	3400
Piston travel, ft/mi	1535	1645
Brake swept area, sq in./ton	196	310

Road & Track *Twin-Test* data sheet (280Z vs Racing 240Z).

racing tyres, while the rear rims were 15in by 15in, with 25 × 13/15 tyres. On the outside, the bodywork was extensively modified, wheel flares being the most obvious feature, though the extended nose and deep front air dam also told you this was no everyday road car.

In some ways, it might come as a surprise to learn that acceleration figures for the racing 240Z were not too far ahead of the production 280Z. For example, the 0–60mph time of the cars was only 2.4 seconds apart, with the production car achieving 9.4 seconds. The standing quarter mile time for the 280Z was 17.3 seconds, while the racer achieved 15.8 seconds, no better than a contemporary 2.7-litre production Porsche Carrera. The real impact of the racing 240Z came with its braking distances and cornering power. The 60–0mph distance was 25ft (8m) less than that for the 280Z, though the pedal effort was much greater, as you would expect. There were no lateral acceleration figures available, as the racing car had never been put on a skid pan, but *Road and Track*'s testers reckoned that the car would generate something like 1.1g, against the proven 0.72g of the 280Z. Top speeds were interesting, too, as the Sharp 240Z reached 147mph (237kph), a clear 20mph (32kph) above that of the production machine.

Once out on the track, the racing 240Z showed its true potential and the huge gap in performance over the production car. Its discs-all-round braking kit meant that braking into corners was very much later than with the 280Z. Rolling acceleration was much greater than the production car, lateral stability far higher, though comfort was something Bob Sharp had not even considered. Getting off to a start was far less smooth in the 240Z, with its much harsher clutch and tighter engine. In fact, the 0–30mph standing start time revealed a great deal about how hard it was to get the car off to a smooth start. This was the one performance figure where the 280Z beat the race car hands down, with 3.2 seconds against 4.4! But to put it into true perspective, the lap times for the two cars tell it all. The production 280Z, after a bit of practice and several adjustments of tyre pressures, managed a very respectable 74.94 seconds on Lime Rock's 1.54 mile (2.48km) track, but the racing 240Z beat that with a startling 57.7 seconds.

NEW MODELS – NEW RACERS

As the product line of Datsun cars improved, there was an ever-pressing quest to improve them still further and to improve their market standing by racing. Consequently, Nissan USA spent quite a lot of money and effort to keep their product at the forefront in sports car racing. Bob Sharp and Don Devendorf were both respected names in the business of racing and both used the talents of Nissan extensively to complement their own development programmes to make Datsun sports cars ever quicker on the track.

By the late 1970s the two prime racing organizations running Datsuns were now fielding 280ZXs, the new shape of Z, and were doing pretty well in both SCCA and IMSA championships. Bob Sharp and the famous actor and driver Paul Newman were winners of the 1979 SCCA Class C Production Sports Car Championship, while Don Devendorf, who had switched loyalties to IMSA in the mid 1970s, had secured the GTU Championship in the same year. This highly successful pair of cars was brought together by *Road and Track* in a fascinating track test, some of which is worth repeating here.

Here were two race team leaders who worked on opposite sides of the United States, Don Devendorf's Electramotive Incorporated being based at El Segundo, in California, while Bob Sharp/PL Newman Racing was based at Wilton, Connecticut. Despite that, and the fact that the SCCA rules allowed the engine of the new 280ZX to go up to 2,818cc, while the IMSA rules still limited the engine to 2.5 litres. So Don Devendorf's car was at a theoretical disadvantage of over 300cc. Both teams had the advantage of support from Nissan USA's competition development department, represented by Dick Roberts, but apart from that, they had their own ideas and pursued their own development paths.

Fundamental differences in the two sporting organizations' regulations meant that there were pretty significant differences between the two 280ZXs. The major differences occurred in the areas of tyres, engines and bodywork. The SCCA rules dictated that wheels had to be of factory standard diameter, so 14in by 7in rims were the limit. IMSA, on the other hand, allowed 16in diameter rims, so 16 × 11s went on the front and 16 × 12s on the rear of Devendorf's car. Common to both cars were the Lockheed vented disc brakes front and rear, Goodyear Blue Streak racing rubber and Gotti wheels. Tyre size differences meant that the Sharp car had much less rubber on the ground than the Electramotive machine. With the larger engine size, Bob Sharp's vehicle produced 300bhp at 6,000rpm, though the Devendorf car came close with 270bhp out of an extra 1,500rpm.

Driven in anger, both cars performed impressively, with the maximum speeds being very close, though the extra rubber on

Proving the Z in Racing and Rallying

the Devendorf IMSA car, together with slightly larger brake discs, which gave an extra 84 sq in (540 sq cm) of surface contact area, meant it could brake much later than the SCCA specification Sharp/Newman motor. Clearly, this gave a potential edge to the Electramotive car, because it could brake later into corners and be assured of being able to stop in an emergency. For example, this car's 60–0mph distance was an astonishing 40ft (12m) shorter than the Sharp car, which required 121ft (37m). Despite all that, on the day of this test, Paul Newman won the bet that stood between the two teams in this informal race test, with a comfortably faster lap time, the additional 30bhp pulling the car through in the end.

By 1981, Don Devendorf had switched from GTU to GTO category with IMSA and really spent the year learning the ropes of his new class. Having done so, he came out charging in 1982, using a 280ZX Turbo, and took six victories to clinch the GTO Championship for that year. Developed out of his earlier 280ZX racer, Don Devendorf's Electramotive company had produced an astounding machine, with 60mph coming up in just 3.8 seconds, zero from 60mph in an incredible 79ft (24m) and a top speed of 145mph (233kph). The downside was 4mpg or 70 litres per 100km!

The arrival on the scene of the VG30-engined V-6 300ZX meant a switch for Bob Sharp Racing to the SCCA's GT-1 category for amateur racing. Paul Newman had a less good season in 1984, but came back in the GT-1 class to win the Championship in 1985 and 1986. Scott Sharp, son of Bob, made it three in a row for the team by taking the Championship in 1987, and then won the SCCA Escort Endurance Championship a year after that with Pete Pombo. Also in 1988, Geoff Brabham – son of Grand Prix World Champion and Formula One car maker Jack Brabham – had swept the board driving a prototype Nissan in the 1988

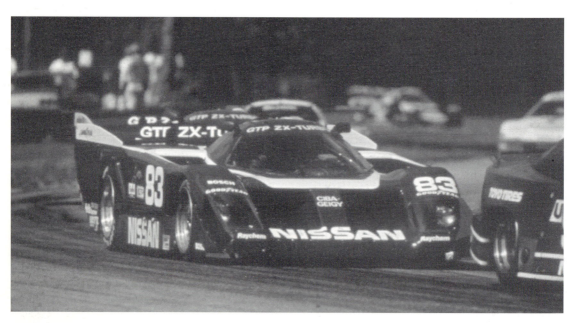

Gary Brabham with the Nissan GTP prototype in the 1989 Sebring 12 Hours Race.

Camel GTP Championship, laying the ground for the next step in Nissan racing successes under the IMSA's GTO category.

RACING THE NEW SHAPE 300ZX

After a tremendously successful run of racing successes in the 1970s and 1980s, Nissan USA continued to support a programme of racing with the new shape 300ZX, announced in 1989. The potential of this magnificent new car was quickly seen by the manufacturer and as the road car was being prepared for announcement to the market, a race version was in development by Electramotive Incorporated, under close works supervision. This was going to be Nissan's winner for the 1990s.

IMSA's new rules required that a GTO car had to have body panels that gave it an identity with its production road-going sibling and the engine had to be in the same position as the catalogue model. Within these regulations Don Devendorf exercised his ingenuity, with factory support, and had an aluminium crankcase and cylinder block casting designed to lighten the overall assembly. Based on the VG-30, the alloy blocked engine looked like the factory standard unit, but reduced to 2.75-litres capacity. It retained twin turbochargers and produced a pretty impressive 690bhp.

The first two 300ZX GTO cars were run by Cunningham Racing and driven by Steve Millen and John Morton in 1989. Despite the power rating and extensive development work done on the engines, the cars were not particularly successful in that first season, turbocharger lag being a problem on twisty circuits. However, for 1990 Nissan took over Electramotive and reorganized it as Nissan Performance Technology Incorporated, located at Vista, California. Nissan meant business and intended to stay at the top of its own sports car racing tree, having had a go at the World Sports Car Championship in 1987 though ultimately suffering, as did most European car makers, from the FISA's indecision and constant rule changes in Group C.

Road America 1991 saw the 300ZX sweep the board.

Proving the Z in Racing and Rallying

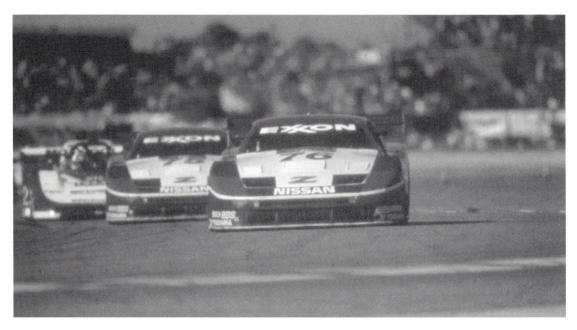

Finally, the 300ZX wins the 1994 Daytona 24 Hour race, the first time a front engined car had won this event since 1977.

In the sports car category the 300ZX Turbo made its racing debut at the Miami Grand Prix, where Steve Millen led the field in the GTO Category from start to finish. It was to be the start of a whole new string of successes for the Z. By the end of 1992, Steve Millen had walked away with the IMSA GTS Championship. The 1993 Sebring 12 Hours Race fell to Millen's 300ZX, a new first for Nissan. Twelve months after that, America's other leading sports car race, perhaps the most prestigious, the Daytona 24 Hours, fell to a 300ZX. It was Nissan's second win at Daytona, a GT Prototype having won in 1992. It was the first time a front engined car had won at Daytona since 1977.

Today, the early Datsun sports cars are also back on the race tracks, though some have hardly left, and classic sports car races regularly see 240Zs, 260Zs and 280Zs taking on all comers. And just to really hammer home the point, the 240Z built by Bob Sharp Racing became a 280Z and was still being raced in front-line events by Jim Fitzgerald a decade later. The Z story goes on.

12 Running and Restoring The Z Series

The first thing you have to decide as a would-be Z owner is which model you want to own. That may, in part, be dictated by the amount of cash you have to spare, for a tatty left hand drive example, in Britain, of any model is likely to cost you a deal less than a pristine restored example in right hand drive form. (The situation of drive position reverses, of course, if you live in the United States or any other left hand drive country.) But unless you want to spend a fair pile on restoration and a certain extent of metal replacement, you might be better off buying a respectable example to begin with. The cash pros and cons are always debatable, but the sheer enjoyment of doing your own work takes a bit of beating, if you have the skill. If you do not have the skill, then the decision is largely made for you, isn't it?

BUYING A Z

It may come as some surprise to the potential classic 'Z' car buyer that the oldest cars are now over twenty-five years old and the youngest of the first generation Zs, the 280 and 280 2+2, are now over a dozen years old. So the fact is that the famous 'metal mite' (otherwise known as rust) will already have attacked your dream car to some degree. The problem facing you is: 'to what degree', for the fact is that the early Z cars are, like many of their European contemporaries, put simply, rust buckets. There is an unfortunate myth that Japanese cars of the late 1960s and early 1970s suffered far worse corrosion than those built in Europe or North America. This really is just a myth – put about by all sorts of people, many who had a vested interest – largely aimed at discrediting Japanese cars before they got too strong a hold on the markets they were invading.

The fact of the matter is that American and European vehicles suffered every bit as badly, and any car you might be thinking of buying that is over twenty years old needs extensive examination. Serious anti-corrosion measures were only just being investigated in the 1970s and there were few anti-corrosion warranties being offered by manufacturers. In much more recent times, it is a very different story and Japanese manufacturers are up there with the rest in their efforts to produce a car that will not rust away in the first five years of its life. It is equally true that if all the world's manufacturers produced cars that did not deteriorate, the market would pretty soon dry up. There is some evidence already that those car makers with a conscience who spend a great deal of time and effort making their vehicles more durable are suffering some consequences, as owners simply replace worn components and hang on to the old car. But the market continues to crave something new and, with new discoveries being made by the industry, the motor industry is not yet ready to die in its feet.

Len Welch's mean green machine – a very original looking 240Z, but it wasn't always like this!

When you have decided which model Z you want to own – the 240Z, the 260Z or the 280Z, or a 2+2 version of either of the later two – you have to go out and find it. How? Well, you could do a great deal worse than join a relevant club and seek out your new steed by that means. The benefit of that line of action is that you could find someone who knows enough about Z models to help you avoid making a major mistake and buying a 'pup'. Even then, of course, it is still possible to miss something, so make sure that, if your chosen car is to remain cherished, you pick one that has the least bodywork problems, for major bodywork repairs can cost a great deal more than mechanical renovation.

As far as the mechanical aspects of a Z are concerned, you will almost certainly find that an old car has mechanical chatters and rattles. They are not necessarily serious problems, but you need to be sure they are not before establishing a buying value for the car. As with so many old cars, there can be rattles as a result of a need for adjustment, or there can be rattles that signify wear. As far as you are able, you need to establish which they are before parting with your money. For example, you can detect a rattle in the top end of the engine that can range from maladjusted tappets to worn little ends on the connecting rods. If you are mechanically minded, you can usually get a feel for what the fault is by carrying out certain checks, but if you are not, you could be buying a lot of trouble without the advice of someone with knowledge of the car type.

Two examples of the kind of problem you could need to identify before buying are piston ring wear and camshaft wear or lack of adjustment. The classic way of identifying whether or not the piston rings are worn (or worse, broken) is to start the engine and when it is warm, leave it running for a few minutes before accelerating the engine speed to about 2,000rpm. You do not need to run the engine too hard, but if you see a puff of blue smoke out of the exhaust as you rev it, then it is almost certain that the rings are worn, which will mean an engine stripdown to rectify the fault.

If there is a light chattering sound from the top end of the engine, it could be a worn camshaft, worn camshaft bearings, or simply bad adjustment. Here, you need to go a little further. Sometimes, you can see enough of a cam lobe through the oil filler to get an idea of what you want to know. If you can, then it is relatively simple to get someone else (perhaps the person who is offering you the car for sale) to turn the engine on the starter key far enough to expose the face of the lobe. If that is misshapen or scored, then camshaft wear is a reasonable suspicion. If, on the other hand, there is no obvious sign of wear, then you need to lift off the cam cover and check the camshaft clearances. Of course, you can never be completely sure with an overhead camshaft engine, because if the clearances seem reasonable and there is still a rattle, it could be that the camshaft bearings are worn.

By and large, unless you find that the engine in your proposed charge is smooth running and reasonably quiet, you must assume that the risk exists for a full engine overhaul. You will almost inevitably find traces of oil leaks here and there and as long as they are not from between the cylinder head and block, or from the clutch housing, you can reckon there may not be too much wrong with the engine. However, oil leaks are irritating things and you should take steps to cure them. If there is one from between the cylinder head and block, suspect a head gasket. Similarly, if there is a leak from the drain hole in the gearbox bell housing, suspect a rear main bearing oil seal. That is a fairly major job. There is a philosophy that says 'leave well alone while you can' and it may be appropriate with your chosen Z, but if you have any doubts about the mechanical condition of the car, either leave it alone or allow for the prospect, and cost, of a major engine overhaul. Use an expert if you follow the latter route.

TRANSMISSION AND FINAL DRIVE

Datsun gearboxes are normally reasonably solid pieces of equipment, giving few problems, so they are not often in need of a huge amount of attention. However, if you find it difficult to select a particular gear, or if one gear 'snicks' as you select it on the move, suspect wear. If there is an oil leak out of the front or back of the gearbox, suspect damaged oil seals or wear. If there is a rumble from the gearbox, there is definitely wear and it is probably the mainshaft that is causing it. If, on the other hand, you find the gearbox quiet in operation, but a bit stiff to put into any gear, then the chances are you have a problem with the clutch. That is a much easier problem to solve. It still means removing the gearbox, but the cost of replacing a clutch plate is much less than that of overhauling a gearbox.

As you go through your car, carefully check the propeller shaft, especially its universal joints. If the car has been driven harshly by a previous owner, especially one who has shunted the car from gear to gear with little regard for the clutch, the propeller shaft may have taken a beating. It is not the end of the world to renovate, but again it can have an influence on the price you pay for the car.

If the propeller shaft is battered, then the chances are that the differential will have had a fair pounding too. However, the Japanese tend to allow for rough road usage in designing their cars, so the final drive is pretty tough. Once more, bearings and oil seals are things to check and any undue movement of differential components will suggest undue wear and cost. The crown wheel is a pretty substantial piece of steel in a Z and the pinion from the propeller shaft is quite long and well supported, but very harsh driving can do some fearful damage and you need to investigate pretty closely to make sure all is well.

If you are inclined to dismantle and restore the rear end of your Z after you have bought it, then there are a few things you need to check carefully, if only to avoid the need to do it all again too soon! One of the first things to examine is the differential torque arrester strap, which is made of rubber and is looped over the nose end of the

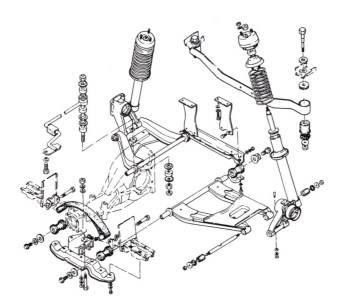

This exploded view shows the rear suspension assembly of the early 240Z.

differential to limit its travel under sharp acceleration or heavy braking and reduce the shock loading transferred into the differential assembly. Quite often, it is lathered in grease or dirt, or a liberal combination of both, so that it is not easy to determine its condition. On an early car, it is almost certain to need replacement, though if it looks sound and is not cracked, it may clean up and be re-usable. But if you are not sure, do not take chances. The universal joints at the back end are also quite important, as those at each end of the drive shafts are most prone to wear and damage through abuse. If you have a chance to jack the car up and check out the rear end, it must be worth doing, because again, the bills can be quite high for replacements and renovation.

BODYWORK

As with most cars of 20 years old or more, regardless of country of origin, bodywork is always suspect and needs very careful scrutiny. An example of this is well demonstrated by early corrosion-protected Porsches, which had zinc-dipped body shells after 1975. They certainly held together for far longer than their immediate predecessors, but were still prone to serious corrosion in little nooks and crannies that had been put there as stiffeners or to hang something on. These proved to be serious water traps, as they were also the nooks and crannies that a fairly quick zinc dip did not necessarily reach, but where moisture, once in, would lie for long periods of time – unless the car was kept in a centrally heated, dehumidified garage, which was a pretty rare situation.

Datsuns probably compare quite well with Alfa Romeos of the same period in their rate of rust. Early Porsches were in much the same category and many a Triumph and MG has finished up in a breaker's yard, assumed to be too rusty for renovation. The usual areas of sills, front wings, wheel arches and floor panels are areas to be examined carefully. It is quite difficult to point the would-be buyer at the worst potential area, as each car is likely to be different. However, the structural members must be examined first, as the extent of rust here can determine whether or not the restoration will be expensive. Starting at the front, the cross member between the front wings and running underneath the radiator should be checked, as must the front suspension mounting plates, where the cross member is attached to the longitudinals below the inner wings. Rust in these areas can be quite serious and may involve the new owner in extensive reconstruction, which is a task that should only be undertaken by someone with the proper equipment and expertise – make no mistake about it, the renovation and replacement of critically structural members is a highly skilled task.

One of the biggest problems with any steel bodied car, and the Datsun is no exception, is that drain holes often block while the car is relatively new and the owner does not realize what to do to unblock them. Water then builds up in undesirable areas and creates the perfect environment for rust. You can often tell the weak areas by doing no more than searching out bubbles in the paintwork. Where there is a bubble, there is a trace of moisture and worse. Carry a magnet during your search for a sound car: just because it looks sound does not mean there is solid metal there – there might be solid filler!

Many a Z has 'lace curtain' front wings and they can often be holed, or very thin, on the vertical surface along the top just below the crease line, all around the wheel arch, down the rear edge adjacent to the join line, below the headlamp mounting and into the sidelamp surround and in the area between the wheel arch and the rear edge. Also check the inner wing, where at its bottom edge, adjoining the longitudinal box section, it can be like a pepper pot! The enclosed boxed area just below the front wings is also an area for suspicion, as this too can rust away from the inside and leave you with flapping wings at the bottom.

Around the area of the front cowl, behind the firewall, is one place where water gathers if drainways are not kept clear. Also, because there are two inspection flaps just ahead of the scuttle, there is another opportunity for

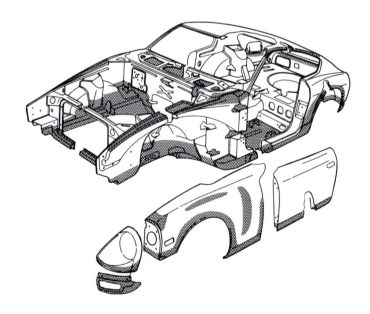

The 'rustbug' hits the shaded areas of panelwork shown in this drawing.

Running and Restoring the Z Series

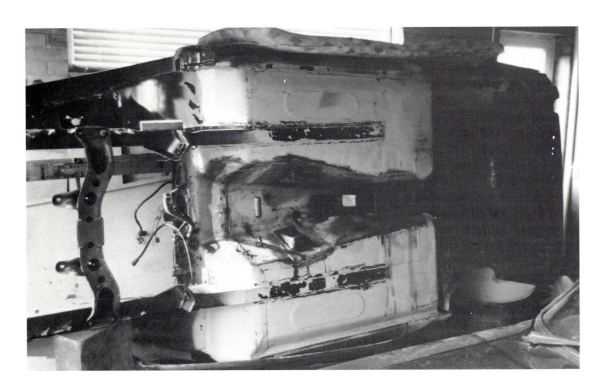

Running and Restoring the Z Series

The rebuild of Len Welch's 240Z, showing progress to a solid body once more (opposite and above).

water to collect. As a result, hidden rust can take place and cause a further need of replacement. Doors also rust, usually from the bottom up and often because water is not drained from the bottom of the doors, so it gathers on the inside. If you want to know how, think of a driver having the window down in a light shower. Water, even only in small amounts, drips down inside the door and when the driver winds up the window again, the water is trapped inside. Rust follows.

Floor panels are always susceptible to rust damage, as the underside is bombarded by flying stones (which chip the underbody protection) and water, often salty in winter. Then, the inside of the floor panel receives wet shoes on the carpeting and if it rains on several days in a row, quite a considerable amount of water can be deposited. That water, again often salty, soaks through the carpet and into the underfelt, lying on the floor and slowly eating its way through the panel from the inside.

You would, of course, have to be very unlucky to find a car with all these faults and rust in all these areas. If you did, you would be wisest just to walk away from it.

BUYING THE BEST CAR FOR YOU

Having walked the tightrope of mechanical disaster and metal mites, if you are still not put off your quest for a Z you have to consider that one suits your needs best. More enthusiasts seem to want the two-seaters, simply because they are the best looking models. The roof line remains clean and pure on a two-seater, whereas the 2+2 has that positively stretched look. However, if you have a small family and still want a Z, your choice is all but made for you: go for a 2+2. As the engines are the same, regardless of body type, so you can still have the fun from the inside.

Another key choice is whether you want the Spartan two-seater, the slightly smoother carburettor-fuelled two-seater or the fuel injected one. The fuel injected version, of

course, is most likely to be left hand drive, but if that does not bother you (or your insurance company in Britain) then press on. The 260Z is in many ways the best choice, as it offers slightly better roadholding than the 240Z and, without the emission gear, a little better performance and is just a bit more pleasant to sit in. The 280Z offers fuel injection, but then you have the problem of having to go to a fuel injection expert when something needs attention. Carburettors are much more easily tuned and dealt with

Len Welch's '10-Day Rebuild' of his 260Z, a car which was not so badly rusted as the 240.

The finished 260Z, looking quite a nice car.

by an amateur than is fuel injection and the real power gain is not all that much. And remember that most road testers seemed to think the longer wheelbase of the 2+2 conferred benefits in terms of handling and roadholding.

You might, on the other hand, prefer a more luxuriously appointed car than the original Z series, but still want it to be a Datsun or Nissan Z. In that case, look the 280ZX, the 280ZX Turbo or the 300ZX. Other than for detail, the interior appointments of these three are not too much different, so your preference comes down to the 2.8 engine, the 2.8 Turbo engine or the V-6 3-litre engine. Again, if you want to retain a link with the original six-cylinder engine, your choice is made and you need look no further. On the other hand, if you want a turbocharged engine, then you need to take a closer look at the 280ZX Turbo, but get someone who knows about turbochargers to look over your choice, to avoid serious damage to your cheque book.

Of course, the V-6 engine of the first series 300ZX offers you the performance of the original 240Z, but you may never notice that, as your environment inside the car is so significantly different, and you are cosseted from the minute you step into the car. It does not even sound the same and everything is so much smoother, you might swear that you were in anything but a Datsun Z.

If, on the other hand, you have decided to fall for the V-6, but want to spent all your life's savings on a truly exhilarating turbocharged car with that Nissan VG-30 engine, you will have to pay for it, for the Z-32 Series 300ZX is still in the band of expensive used exotics and you will almost certainly pay nearly as much as for a good used Porsche 944 Turbo. The advantage, if you can afford it, is that you are not buying rust problems you cannot do anything about and you are buying a state of the art car with scintillating performance at a fraction of the price of many of its adversaries – and it may well prove to be every bit as durable.

Whatever your preference, think hard about your impending purchase, decide clearly what you want to use it for when you have bought it, then go and buy it, preferably with the advice of someone who knows something about Z models. And above all – enjoy it.

Running and Restoring the Z Series

Z CLUBS OF THE WORLD

Around the world, there are many clubs catering for the interests of Datsun/Nissan enthusiasts and owners. This list is offered as an indication of the clubs that exist and where they are, but its accuracy cannot be guaranteed, nor does it claim to be absolutely complete. But it is a start.

Australia
DATSUN SPORTS OWNERS CLUB
PO Box 402
South Yarra
Victoria 3141

HUNTER VALLEY Z CAR CLUB
Mark St Claire
21 Morris Street
Birmingham Gardens
New South Wales 2287

ACT AND DISTRICT Z CLUB
Alan Beltrami
PO Box 385
Queanbeyan
New South Wales 2620

Canada
Z CLUB OF CALGARY
Blaine Worger
621 11 Avenue NE
Calgary
Alberta T2E 0Z8

Z CLUB OF CANADA
Milton Joneson
7933 178th Street
Edmonton
Alberta T5T 1L3

France
Z DRIVERS' CLUB
P. Braun
10 Residencechateau Bois
62131 Vaudricourt

Germany
Z AND ZX CLUB OF GERMANY
Fred Schuster-Orth
Messeplatz 4
D8000 München 02

Great Britain
THE CLASSIC Z REGISTER
Lynne Godber
Thistledown
Kentsborough
Middle Wallop
Stockbridge
Hampshire SO20 8DZ

Z CLUB OF GREAT BRITAIN
Steve Burns
11 Greenford Gardens
Greenford
Middlesex UB6 9LY

Netherlands
Z AND ZX CLUB OF HOLLAND
Guss van der Boll
De Sav
Lohmanlaan 45
NL2566 Den Haag

New Zealand
THE DATSUN Z CLUB AND Z CLUB OF
NEW ZEALAND
Graham Collins
PO Box 24-176
Royal Oak
Auckland 6

Norway
NISSAN SPORTING
Roy Arne Gulbrandsen
Boks 30
2092 Minnesund

Sweden
Z CLUB OF SWEDEN
Johann Lundin
MJ Olner Backen
172.48 Sundbyberg

Switzerland
Z CLUB OF SWITZERLAND
Albin Henseler
Kaserei
6042 Dietwil

United States
THE Z CLUB OF ARIZONA
Dale Sheetz
PO Box 7006
Phoenix
AZ 85011

GROUP Z
Roger Patawaran
PO 10497
Santa Ana
CA 92711

Z OWNERS OF NORTHERN
CALIFORNIA
Mike Voss
PO Box 272782
Concord
CA 94527

Z CLUB OF COLORADO
Eva Noworytta
PO Box 13455
Denver
CO 80201

WINDY CITY Z CLUB
Steve Appelbaum
PO Box 6009
Evanston
IL 60204

Z CLUB OF NEW ENGLAND
Frank Ricci
PO Box 75
Framingham
MA 01701

GATEWAY Z CLUB INC
Michael Breeding
PO Box 220023
St Louis
MO 63122

Z CLUB OF ALBUQUERQUE
Rocky Drebber
3224 Florida NE
Albuquerque
NM 87110

Z CLUB OF CENTRAL OHIO
Bill Antoniak
PO Box 19808
Columbus
OH 43219

NORTH COAST CAR CLUB
Steve Kinosh
13a 826 Elsette
Cleveland
OH 44135

PIEDMONT Z CLUB
Keith Vaughn
Route 1
PO Box 261
Easly
SC 29642

WHITE ROSE Z AND ZX CLUB
Steve Brown
118 East King Street
York
PA 17404

Z CLUB OF HOUSTON
Vanessa Jacobs
PO Box 66551
Houston
TX 77266

Z CLUB OF TEXAS
Steve Vorenkamp
PO Box 937
Keller
TX 76248

Further Reading

As might be expected, a great deal of research material had to be found for the preparation of this book. It was made the more difficult because of the lack of a substantial archive at Nissan Motor (GB) Limited, since that company was formed in 1992 to take over the total distribution and promotion of Nissan cars and little or nothing was handed over from its predecessor.

The list of material and publications given here is aimed at providing the reader or enthusiast with more material. Some, but notably not all, was used in the research for this book.

- Datsun/Nissan brochures, 1969 to 1994
- Various documents from the Nissan Archive in Tokyo
- Documents loaned by Nissan Motor (GB) Limited
- Documents and material provided by Nissan Europe NV
- Service and Parts Manuals published by Nissan Motor Company
- A Study Paper on Fibre Reinforced Plastics (Tokyo University)
- The Archive of L.E. Welch Esq
- The Datsun 240Z 1970–73 (Brooklands Books)
- 'Road and Track' on Datsun Z 1970–1983 (Brooklands Books)
- Datsun 280Z and ZX (Brooklands Books)
- Nissan 300ZX (MRP)
- How to Restore Your Datsun Z-Car (Fisher Books)
- How to Rebuild Your Datsun/Nissan OHC Engine (Fisher Books)
- Various issues of Autocar and Motor
- Various issues of Road and Track
- Various issues of Autosport
- Various issues of Car and Driver
- Material sourced from Pook's Motor Books, Leicester

Index

Page numbers in italics denote illustrations.

Aaltonen, Rauno 68, *70*, 167, 168
AC Cobra 41
AC428 68
Acropolis Rally 167
Act for Import and Export of Goods 15
Alfa Romeo 6c3000CM 123, *123*
Alfa Romeo SpA 42
Alfa Romeo 1750 Spyder 68
Alfa Romeo 1750GTV 54, 127
Alfa Romeo 2000GTV 54, 68, 71
Alfa Romeo Alfetta GTV2000 131, *133*
Alfa Romeo Alfetta GTV6 2.5 136, 137, *137*, 156 162
Alfa Romeo Giulia Spyder 34, *37*
Alfa Romeo GT *44*
Alfa Romeo GTV 47, *128*, 155
Allied Control Commission 16, 17, 18, 23
Aoyama, Rokuro 11
armoured car, Type 41 12
Asahara, Genshichi 21, 22
Aston Martin Limited 41
Aston Martin DB6 *50*, 51
Aston Martin DB7 113, *113*, 144
Aston Martin DBS 68, *69*, 152
Aston Martin V-8 engine 124
Austin A40 Somerset 22, 23
Austin A50 Cambridge 24, *24*, 26
Austin Healey 73, 96
Austin Healey 100 28
Austin Healey 3000 42, *44*, 45, 46, 55, 128
Austin Motor Company 22, 23, 24
Austin Seven 14
Australia 13, 14, 19, 28, 41, 69, 109, 166
Australian Mobilgas Trial 28, 165
Autocar magazine 156, 157, 163
Autosport magazine 149
Ayukawa, Yoshisuke 13
Azabu-Hiroo 11, 12

Bedard, Patrick 155
BMW 42
BMW 503 33

BMW 530i 84
Bob Sharp Datsun 170
Bob Sharp Racing 174, 176
Bolster, John 149
Borg Warner automatic transmission 105
Bosch L-Jetronic fuel injection 84, 85, 90
Brabham, Gary *174*
Brabham, Geoff 174
Britain 41, 60, 68, 69, 70, 72, 75, 79, 82, 107, 122, 126, 128, 141, 146, 156, 163, 164, 166, 177
British Leyland 42
British Motor Corporation 22, 25, 27, 55, 128
Brock, Peter 73

California 85, 90, 153, 164
Canada 58, 60, 82
Car and Driver magazine 26, 149, 150, 152
Car Life magazine 62, 146
Chapman strut 47, 48, 49
Chevrolet Corvair 27
Chevrolet Corvette 28, 30, 54, *54*, 100, 162, 163, *163*
Chevrolet Division, GMC 14
Chicago Auto Show 108
China 15
Classic & Sports Car magazine 152
Costello MGB GT V-8 131
Cray Supercomputer 109, 114

DAT 11
Datson 12
Datsun 210 Sedan 28, 165
Datsun 240Z 26, 27, 41, *45*, 52, 54, 57, 59, 60, *60*, 61, *61*, 62, 65, 68, *70*, 71, 72, 75, 76, 76, 77, 83, 87, 90, 104, 126, 165, 167, *168*, *171*, 178, *178*
and its adversaries 125
brakes 74
designs 43, *43*

interior 51, 52, *52*
rear suspension 47, 180
restoration *182*, *183*
road test data 147, 148
Datsun 260Z 76, 77, 78, *79*, 83, 133, 136, 149, 152, 178, 185
and its adversaries 130
restoration *184*, *185*
road test data 150, 151
Datsun 260Z 2 + 2 79, 80, *80*, 81, 82
and its adversaries 132
Datsun 280Z 83, 84, *84*, 86, *87*, 87, 88, 89, 90, 153, 154, 172, 177, 178
and its adversaries 134
road test data 154
Datsun 280Z 2 + 2 86, *86*, 89, *89*, 90, *91*, 92, 177, 178
and its adversaries 135
Datsun 280ZX 91, 93, *93*, 94, 97, 98, *98*, 101, 103, 104, 136, 155, 156, 162, 185
and its adversaries 138
rear suspension, 95
road wheels 97
Datsun 280ZX 2 + 2 96
and its adversaries 139
road test data 157
Datsun 280ZX Targa 99, 100, *100*
Datsun 280ZX Turbo 99, 174, 185
and its adversaries 140
Datsun 310 Bluebird 25, 26, *26*, 27
Datsun 1600 165, *165*
Datsun 1600SSS Bluebird 68, 166
Datsun bodywork 180, 181, *181*
Datsun CD-3 19, *19*, 20
Datsun clubs 178, 186
Datsun CSP311 32, 33, 37, 38
Datsun Fairlady Z 42, 58, *59*, 65
Datsun gearbox examination 179
Datsun L20 engine 58
Datsun L24 engine 46, 47, 49, *49*, 58, 76, 85
Datsun L26 engine 85
Datsun L28 engine 84, 85, *85*,

Index

97, 101, 185
Datsun L28ET engine 100
Datsun Project HS30 45, 46, 55
Datsun Roadster 14
Datsun S-20 engine 65
Datsun S211 28, *29*, 29, 31
Datsun SP200 Series 42
Datsun SP212 29
Datsun SP310 30, *30*, 31, *31*, 34, 55
Datsun SP311 32, 33, *33*, 34, 35, 55
Datsun SR300 Series 42
Datsun SR311 35, 36, *36*, 147, 148
Datsun SR311 Hardtop 37, *37*
Datsun transmission examination 179
Datsun USA 166, 169
Datsun VG30 engine 185
Datsun Violet 70
Datsun Z432 46
Datsun Z-432R 46, 65, *66*, 67, 76
Datsun/Graham Paige Model 70 14, *14*
Datsun/Nissan sports cars: numbers built 160
Daytona 24 Hours Race 176, *176*
Den, Kenjiro 11
Devendorf, Don 173, 174, 175, *175*
Durasteel 110

East African Safari Ralley 68, 166, 167, 168, 169, 170, *170*
Easter, Paul 167, 168
Edwards Deming Prize 25
Electramotive Corporation 174, 175
Enever, Sydney 146
engine examination 178
Experimental Safety Vehicle 27
Extraordinary Funds Adjustment Act 15

Ferrari 250GTO 43, 45, *45*, 47
Ferrari 250MM 123, *124*
Ferrari 308GTB 133, *133*
Ferrari 348TB 113, *113*, 144
Ferrari 365GTB 68
Ferrari Daytona 365GTB 155, *155*
Ferrari Dino 246GT 62, 68, 71, 72, 127

Ferrari Mondial 137, *141*
Fiat 19
Fiat 124 Coupé 47
Fitzgerald, Jim 176
Ford Capri 58
Ford Capri 2600 71, *126*, 127, 128, 129
Ford Capri 2800 131
Ford Capri 3000GXL 82
Ford, Henry 25
Ford Motor Company 14, 25
Ford Mustang 27, 41, 82
Ford of Britain 58
Ford of Germany 58
Ford Thunderbird 96
Frankfurt Motor Show 98, 156

Garrett AiResearch turbo 98
Giugaro, Giorgetto 133, 136
glass-fibre 25
Goertz, Albrecht 31, 32, *32*, 33, 37, 38, 39, 41, 42, 43, 55, 83, 97, 108, 109, 124
Gorham, William R 12
Graham Paige 14, 19, 21
Graham Paige Model 629 14

Hagborn, Gunnar 167
Harris, Peter 95
Hashimoto, Masujiro 11, 12
Havashi, Atushi 25
Herrmann, Edgar 167, 168
HICAS rear suspension 114, *114*
Hiroshima 17
Hokkaido 10
Holland 69
Honshu 10
Hutton, Ray 163

Imperial Japanese Army 16
Imperial Japanese Navy 15

Jaguar E-Type 39, 40, *40*, *50*, 51, 68, 152
Jaguar XJ-S 108
Japan 10, 11, 16, 18, 19, 23, 27, 55, 79, 101, 110
Japan Motor Vehicle Distribution Co. 17
Japanese GT Grand Prix 169
Jidosha Seizo DAT Company 12, 13, 101
Jidosha Seizo Kabushiki Kaisha Co. 13, 14

Jitsuyo Jidosha 12

Kallstrom, Harry 167
Kaminari 164
Kenya 2000 Rally 169
Korea 16, 18
Kume, Yutaka 109
Kwaishinsha 11, *11*, 12
Kyushu 10

Lancia Fulvia 1600HF 167
Lime Rock Park, Connecticut 170, 173
London Motor Show 30, 68
Longbridge 21, 22, 23
Lotus Cars 25
Lotus Eclat 136
Lotus Elan 39, 57, *57*
Lotus Elite 136
Lotus Esprit 136, *137*, 155
Lotus Esprit 2.2 Turbo 141

MacPherson 48, *48*, 97
Manchuria 15
Marcos 3-Litre 71, 126, 127
Maserati Ghibli 68
Maserati Merak 137, 141
Mason, George 22
Mazda RX-7 144
Mercedes Benz 42
Mercedes Benz 280SL 46
Mercedes Benz 500SEC 144
MG Midget 18, *18*, 64
MGA 62, 147
MGB 30, 34, 35, 46, *46*, 48, 49, 54, 64, 73, *73*, 126, 129, 146
MGB GT 47
MGB GT V-8 129, 131
MGC 42, 45, 55, *56*, 62
Miles, John 156
Millen, Steve 175
Ministry of Commerce and Industry 16
Mitsubishi Starion 141
Monte Carlo Rally 167, 170
Morton, John 166, 175
Motor magazine 149
Motor Sport magazine 166
Motor Trend magazine 152, 156, 158, 162
Motorsports International 163
Munitions Industry Mobilization Act 15
Nagasaki 17

191

Index

Nash Metropolitan 22, *22*
Nash Rambler 22
Newman, Paul 169, *171*, 173, 174
Nihon Sangyo 14
Nissan 12, 13, 14, 15, 16, 17, 18, 19, 21, 23, 24, 25, 28, 29, 37, 42, 55, 58, 71, 73, 75, 76, 83, 94, 85, 89, 90, 97, 101, 105, 111, 113, 126, 165
Nissan 300ZX 101, *102*, 103, *107*, 141, 144, 164, *164*, 174, *176*, 185
  and its adversaries 142
  Convertible 120
  multi-link suspension *114*, 114
Nissan 300ZX (Z32) 109, *109*, 111, *111*, 112, 185
Nissan 300ZX 2 + 2 105, 118, 122, *122*
  and its adversaries 143
Nissan 300ZX 2 + 2 (Z32) and adversaries 145
Nissan 300ZX 2 + 2 Turbo 119
Nissan 300ZX 2-Seater 116
Nissan 300ZX 2-Seat Turbo 115
Nissan 300ZX SE *108*
Nissan 300ZX Turbo 101, 104
Nissan 300ZX Turbo 2-Seater 106
Nissan 300ZX Turbo SR-71 164
Nissan automatic transmission 117
Nissan Cedric *26*, 26
Nissan Fairlady 300ZR 108, 111
Nissan Fairlady Convertible 120, 121, *121*
Nissan GTP Prototype *174*
Nissan Performance Technology Inc. 175
Nissan Project HS30 110
Nissan Project UZ 109, 110
Nissan United States 95
Nissan USA 175
Nissan VG-30 engine 101, 104, *104*, 107
Nissan VG30DE engine 108
Nissan VG30-DETT engine 117, *117*
Nissan/Austin A50 Station Wagon 25, *25*

Opal GT 47, 64, *64*, 65
Oppama plant 27
Osaka 12, 14
Pombo, Pete 174

Pontiac Firebird Turbo SS 100
Porsche 911 46, 68, 71, 82, 84, 110, 113, 124, 127, 136
Porsche 924 131, *131*, 136, *136*, 156, 162
Porsche 924 Turbo 100
Porsche 928 108, *109*, 162
Porsche 944 107, *107*, 108, 110, 144, 163
Porsche 944 Turbo 110, 113, 185
Porsche Carrera 2.7 172
Porsche, Dr Ferdinand 22
Portugese Montana Rally 169
Price Control Ordinance 15
Prince Gloria 27
Prince Motor Company 27
Prince Skyline *26*, 27
production numbers: Datsun/Nissan sports cars 160
programme 901 110

RAC International Rally 68, 167, 170
Reliant Scimitar GTE 57, 71, *71*, 126, 127, 129
Rising Sun 14
Road & Track magazine 26, 147, 148, 149, 151, 152, 153, 154, 162, 170, 172, 173
Road Test magazine 74
road test data 240Z/280Z 172
Roberts, Dick 173

Schuller, Hans 167, 168
Sebring 12 Hours Race 176
Sharp, Bob 166, 169, *171*, 173
Sharp, Scott 174
Shelby, Carroll 41
Shikoku 10
Singapore Grand Prix GT Races 169
Singer 19
Southern Cross Rally 169
Sports Car Club of America 60, 60, 69, 167, 169
Sports Car Graphic magazine 62, 63, 64, 65, 146
Strike, 100-Day 23
Studebaker Golden Hawk 28

Takeuchi, Metaro 11
Tanzania 1000 Mile Ralley 170
Thousand Lakes Rally 167
Tobata Imono Automobile Division 13

Tobata Imono Company 13, 14, 101
Tokyo 10, 11, 54
Tokyo Motor Show 30, 31, 37, 82, 121
Tokyo University 25
Towns, William 152
Toyota 17
Toyota 2000GT 33, 38, *38*, 39, *40*, 40, 41, 67
Toyota Celica Supra 3-Litre 156
Toyota Celica vs Datsun 280ZX 158
Toyota Supra 3-Litre 141
Triumph GT6 47, 48, 57, 68, 126, 128, *129*
Triumph Stag 136
Triumph TR4 *34*, 35
Triumph TR5/250 *56*, 57, 62, 126, 128,
Triumph TR6 57, 68, *69*, 71, *73*, 129
trucks 12, 16, 17
TVR 3000M 131
TVR Tuscan 57
TVR Vixen 68, 71, 126, 127, 129

United Nations 16
United States 10, 11, 19, 22, 27, 28, 29, 31, 33, 35, 41, 42, 46, 48, 49, 51, 54, 55, 58, 61, 72, 75, 76, 77, 79, 82, 83, 85, 93, 98, 100, 107, 109, 110, 121, 126, 127, 128, 131, 144, 153, 156, 165, 166, 169, 177
US Federal Motor Vehicle Safety Standards 35, 61, 83

Vauxhall Cresta 26
Volkswagen 22

Wankel, Dr Felix 144
Welch, Len 32
Welsh Rally 169
World War One 12
World War Two 16, 64

Yamaha 38, 39, *39*, 41
Yamaoka, Shigeyuki 110
Yokohama 13, 14, 16, 22, 23, 24

Zama plant 27